Mastodon-Bearing Springs and Late Quaternary Geochronology of the Lower Pomme de Terre Valley, Missouri

Figure 1. (Frontispiece) Aerial photograph of the Pomme de Terre Valley in the vicinity of the Breshears Valley, an abandoned incised meander. 1) Boney Spring, 2) Trolinger Spring, 3) Jones Spring, 4) Kirby Spring, 5) Koch Spring, 6) Phillips Spring, and 7) Rodgers Shelter.

Mastodon-Bearing Springs and Late Quaternary Geochronology of the Lower Pomme de Terre Valley, Missouri

C. Vance Haynes, Jr.
Departments of Anthropology and Geophysics
University of Arizona
Tucson, Arizona 85721

SPECIAL PAPER
204

Published by The Geological Society of America, Inc.
3300 Penrose Place, P.O. Box 9140, Boulder, Colorado 80301

Printed in U.S.A.

Library of Congress Cataloging in Publication Data

Haynes, C. Vance (Caleb Vance), 1928-
Mastodon-bearing springs and late Quaternary
geochronology of the lower Pomme de Terre Valley,
Missouri.

(Special paper / Geological Society of America ;
204)
Bibliography: p.
1. Geology, Stratigraphic—Quaternary.
2. Springs—Missouri—Pomme de Terre River Valley.
3. Radioactive dating. 4. Geology—Missouri—
Pomme de Terre River Valley. I. Title.
II. Series: Special paper (Geological Society of
America) ; 204.
QE696.H38 1985 551.7'9 85-12546
ISBN 0-8137-2204-7

Contents

List of Figures

List of Tables

Preface

In September 1965, I received a letter from W. Raymond Wood, then Director, River Basin Surveys, University of Missouri, asking if I would be interested in conducting geological work in conjunction with archaeological excavations to be conducted at a bluff site called Rodgers Shelter. Aside from the intriguing potential for Paleoindian occupation of the shelter, Wood offered more "bait" by saying " . . . about a mile south of the shelter, and in the same river valley, is the presumed locality of the Koch mastodon on the Pomme de Terre River." The potential for finding evidence for pre-Clovis Paleoindians was too good to pass up, but of further interest to me was the opportunity to work out the geochronology of the alluvial sequence in the subhumid unglaciated Ozark Highlands, well away from the arid and semiarid Southwest where most of my work on alluvial stratigraphy had been done (Haynes, 1968). The sites also happened to be near the childhood homes of a long-standing close friend and his wife who had impressed me with the surface collections of artifacts they had made there. So, in July 1966, Jerry and Carolyn Shelton and their two boys joined my wife, Taffy, daughter, Lisa, and me on a trip to the northern Ozarks where we joined Ray Wood and R. B. McMillan.

Ray Wood and I also coaxed Peter J. Mehringer, Jr., into looking for fossil pollen there. Preferring spring bogs to alluvium, he cored Boney Spring and discovered, in addition to pollen, bone and tusk 4 m below the surface (Mehringer and others, 1968). The following summer we discovered the bone beds and peat deposits of both Boney and Trolinger springs. By this time Bruce McMillan, who was conducting the excavations at Rodgers Shelter for his Ph.D. dissertation, had already found abundant evidence of Paleoindian occupation as long ago as 10,500 B.P. (McMillan, 1971). Later, he established the true location of Koch Spring Spring which we excavated in 1971. We examined Jones Spring in 1966 and tested it in 1971 when the lower peat was discovered.

James E. King took over the fossil pollen investigations in 1968, and with Bruce McMillan in 1973 we discovered the Phillips Spring site while backhoe trenching in the spring deposits. The vertebrate paleontological studies of Trolinger and Boney springs were under the direction of Everett H. Lindsay and Jeffrey J. Saunders, who studied the fauna of Boney Spring and conducted the paleontological excavations at both Jones and Trolinger Springs from 1971 to 1977 (Saunders, 1977a, 1977b), while Stephen Chomko (1976) and Marvin Kay (1980) excavated Phillips Spring.

Up to 1976 these investigations had been financially supported by the National Park Service and the National Science Foundation, but thereafter the U.S. Army Corps of Engineers has supported most of the work by contract with the University of Missouri, under the direction of first W. R. Wood and later Donna C. Roper.

In many ways, other than just scientific, my 14 years of participation in these investigations have been exceptionally rewarding. Close ties of friendship have been established with both the project's participants and the people of the area. And the avocational preservation of a pioneer log house has given me and my family a keen interest in the history of the area and some new roots. Even though subsistence farming is a thing of the past (West, 1945), the Breshears Valley is still a good example of rural America.

My only regret is the loss of the primitively beautiful valley to Truman Lake and the displacement of the old folks living there. They say the dam is for flood control (U.S. Army Corps of Engineers, 1975) and perhaps we would not have discovered the 100,000 years of pre-history were it not for "the project," but why create a permanent flood over 55,600 acres to prevent occasional flooding of about 30,000 acres! Perhaps a more practical approach would have been to restrict human use of the lower terraces (T-0 and T-1) to agriculture and recreational use. Senior residents of these floodplains could have lived out their lives there on a voluntary basis with the understanding that they did so at their own risk, and subsequent generations could have been prohibited from living there. Or they could have been required to move at government expense to higher ground, but still permitted to own and operate their family farms. This plan would have cost the taxpayers less, preserved the dignity and domestic tranquility of the older generation, and allowed the river to "do its thing."

Acknowledgments

This work has been supported by the National Science Foundation with grants to University of Arizona, University of Missouri, Southern Methodist University, and Illinois State Museum, and by the U.S. Army Corps of Engineers, Kansas City District, with contracts to the Illinois State Museum and the University of Missouri. Personnel of the Corps of Engineers, H.S. Truman Reservoir Project, who were especially helpful are Col. Griffith, Melvin A. Johnson, Roberta Comstock, Mary Lucido, Cecil Ryder, and Edward Elmoe. The Missouri Department of National Resources was very helpful in providing information and assistance in coring operations and completion of wells at Jones and Trolinger Springs.

Lasting friendships have been made with the many colleagues who have participated in the project and made my work possible. This includes Stanley A. Ahler, William H. Allen, David A. Baerreis, G. R. Brakenridge, Carl H. Chapman, Stephen A. Chomko, Donald L. Johnson, Marvin Kay, Gerald Kelso, Francis B. King, James E. King, Walter E. Klippel, Everett H. Lindsay, R. Bruce McMillan, Peter J. Mehringer, Jr., Michael V. Miller, Paul W. Parmalee, J. R. Purdue, Christine K. Robinson, Donna C. Roper, Jeffrey J. Saunders, Ernst A. Stadler, Donna Watson Stegner, Ronald Ward, and W. Raymond Wood, who had the foresight to initiate the interdisciplinary studies of which this is only one part.

A major contribution of this work is the more than 140 stratigraphically controlled radiocarbon dates provided through cooperation with E. Mott Davis and Sam Valastro, Texas Memorial Museum; James B Griffin, University of Michigan; Herbert Haas, Southern Methodist University; Austin Long, University of Arizona; Minze Stuiver, University of Washington; and E. R. Taylor, University of California, Riverside.

During the past decade and a half I have been assisted in one way or another by many citizens of Benton and Hickory Counties, all of whom I consider close friends. Unfortunately, space restrictions allow only a few to be mentioned specifically. Jack Rodgers always enthusiastically supported archaeological excavation of his rock shelter and helped several visitors, uninitiated to the site, extract vehicles that became stuck on wet days. For more than a decade Jesse Kaufman provided very skilled and reliable backhoe service. Homer Routh and Edsel Breshears did equally good work for us after Jess' retirement. Long time residents of the Avery area, Jake and Iva Bird, Iven and Goldina Trolinger, Ray and Lucille Sherman, Bob Kennedy, G. T. and Emma Blackwell, Francis Kirby, Harold Pippins, and Imogene Stuart Nazar provided a wealth of information on the history of the Breshears Valley and the locations of springs and archeological sites. Their patience with the many --ologists excavating on their former farms is sincerely appreciated.

Others who helped in one way or another include Tommy Blackwell, Roy Butler, Robert Drake, Jim Henderson, David and Nancy Herbert, Chip Hyzer, Don Hotolling, Joe Kennedy, Lawrence Kennedy, Kenneth Lasswell, Henry and Kathleen Liedke, Donna Mann, John Salee, and Dillon Tipton.

Over the 14 field seasons in the Ozark Highland I have enjoyed working with many archaeological field crews and graduate students pursuing research in one discipline or another. Their enthusiasm and interest is a rewarding experience. A special word of appreciation is due Alison Habel for her expert drafting and Doris Sample for typing and preparation of the manuscript. Taffy and Lisa Haynes provided constant encouragement and lasting companionship throughout the project.

Preliminary versions of the typescript were read by G. R. Brakenridge, Donald L. Johnson, Marvin Kay, James E. King, Paul S. Martin, Jeffrey J. Saunders, and Michael R. Waters. The manuscript has been significantly improved by the comments and suggestions of the GSA reviewers W. R. Farrand, J. H. Hartshorn, and an anonymous person.

Geological Society of America
Special Paper 204
1985

Mastodon-Bearing Springs and Late Quaternary Geochronology of the Lower Pomme de Terre Valley, Missouri

ABSTRACT

Fourteen consecutive field seasons of scientific excavations of alluvial deposits of the lower Pomme de Terre Valley, southeastern Missouri, have provided a radiocarbon-dated (154 analyses) chronostratigraphic sequence in four terraces reflecting five major episodes of aggradation and degradation. Probably all of Wisconsinan time and the Holocene is represented. A sixth alluvial deposit is probably Illinoian and possibly older Pleistocene alluvium. Spring deposits within the terraces contain bone beds associated with peat lenses containing pollen, plant, and beetle remains, all of which have provided paleoecological data.

The springs appear to have erupted initially, in response to either tectonic disturbance or hydrostatic pressure near the peak of a glacial maximum. Subsequent episodes may correlate with later maxima. Peat deposits correlate with interstadials on the bases of fauna, flora, and radiocarbon dating. Gravel deposits of spring conduits contain bones and plant remains that are considerably older than overlying peat deposits. Some of these older elements have been redeposited in younger spring deposits via spring action. Other evidence suggests that some animals may have fallen into springs and intruded older deposits. A developmental sequence for the origin of concentric spring deposits in alluvium is proposed.

None of the six spring deposits excavated contained any associated archaeological remains older than 10,500 B.P., which marks the first appearance of artifacts in the stratigraphic record at Rodgers Shelter. Extinction of the Pleistocene big-game animals occurred sometime between then and 16,000 B.P.

A major episode of degradation occurring between 10,500 and 13,000 B.P. probably coincides with similar episodes occurring approximately 12,000 B.P. on many if not most streams in the United States. With some exceptions, subsequent epicycles of cutting and filling appear to correlate generally over the same area.

INTRODUCTION

In the valley of the lower Pomme de Terre River, Benton and Hickory Counties, Missouri, an abandoned incised meander forms an isolated valley in which the Henry and Alexander Breshears families made permanent settlement in 1838 (Fig. 1). Two years later they hosted Albert Koch, a German emigrant who collected fossil bones for his museum in St. Louis. He extracted mastodon bones from a spring less than 2 miles south of the abandoned meander, now called the Breshears Valley. Some of the bones made up one of the finest mounted specimens of *Mammut americanum* known. It may be seen today in the British Museum where it has been since 1844 (McMillan, 1976a).

Koch also found stone projectile points in the spring which convinced him that early man had a hand in the animal's demise. These factors, plus the discovery of artifacts in a stratified rock shelter on the nearby Rodgers farm, led to scientific investigations which began in 1963 (Wood and McMillan, 1976). Since then the most complete record of geologic-climatic events in the mid-continent for the last 100,000 years has been determined. This has been accomplished over 15 field seasons by scientifically controlled excavations at the archaeological and paleontological sites. Skilled teams from the University of Missouri, the Illinois State Museum, and the University of Arizona have been responsible for the scientific investigations supported by funds from the National Science Foundation. Additional support from the U.S. Army Corps of Engineers allowed backhoe trenching and core sampling of floodplain alluvium throughout the study area (Fig.

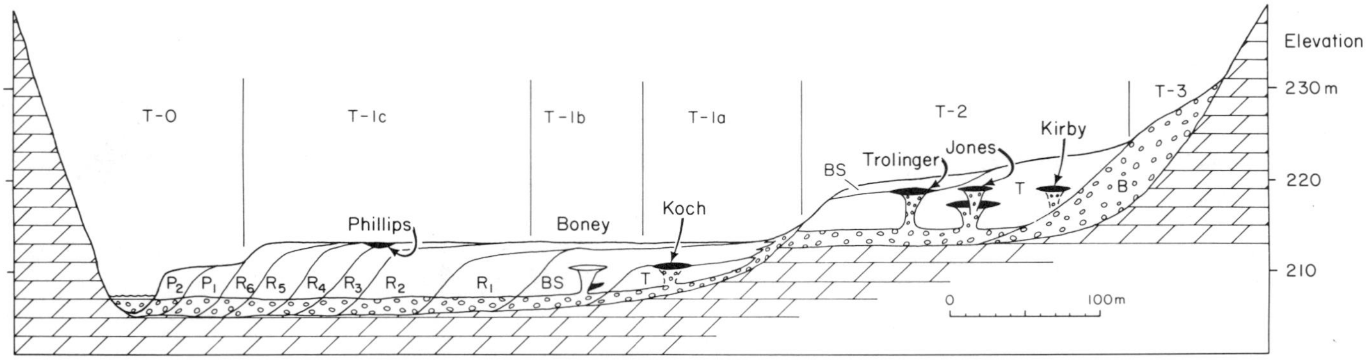

Figure 3. Generalized geological cross section of the Pomme de Terre Valley showing the relative stratigraphic positions of spring-laid peat lenses, alluvial terraces, and alluvial deposits: Breshears Valley Fm. (B), Trolinger Springs Fm. (T), Boney Springs Fm. (BS), Rodgers Shelter Fm. ($R_{1\text{-}6}$), and Pippins Cemetery Fm. ($P_{1\text{-}2}$).

2—in pocket inside back cover). This proved invaluable in deciphering the complex and usually subtle stratigraphy (Fig. 3). The stratigraphic framework of both riverine sediments and intricate spring deposits could not have been accurately defined without the use of stratigraphic trenches, because natural exposures are few and the succession of repeated lithologies makes it essential that contact relationships be observed in adequately exposed and prepared sections. It could not have been done by core sampling alone, which was most useful in determining the subsurface extent of stratigraphic units only after the framework had been worked out from the excavations.

The purpose of this paper is to present the stratigraphic framework from which the various fossil and archaeological finds can be related to one another and to radiocarbon dates and other samples. These relationships provide the basis for the reconstruction of the fluvial history of the Pomme de Terre River (Fig. 4a) and paleoenvironmental changes during the last 100,000 years (the late Quaternary Period).

ALLUVIAL DEPOSITS

The important archaeological sites at Rodgers Shelter and Phillips Spring and paleontological sites at Koch, Boney, Trolinger, and Jones Springs all occur in deposits of three alluvial terraces (T-0, T-1, and T-2) which are further subdivided into alluvial cut-and-fill events (Fig. 3) dated by radiocarbon analysis. The stratigraphic framework as presently understood represents a significant revision of Haynes (1976, 1978) and Brakenridge (1979, 1981). In some places the terrace surfaces are so uneven and close in elevation that they cannot be distinguished on the basis of elevation alone, especially if the contact between them has a low-angle dip. The deposits of T-2 are best represented in the Breshears Valley, an abandoned meander (Figs. 1 and 2) that contains three of the springs important to this study (Fig. 4c). A fourth terrace (T-3) of possible pre-Sangamonian age (Fig. 4b) is mainly a strath of oxidized chert gravels with a strong red, relict soil (Haynes, 1977). These gravels are here named the Breshears Valley Formation (Fig. 3). The T-3 terrace has not been studied in detail, but its point bar deposits probably represent the last time the river meandered on the scale indicated by the incised meanders of the Pomme de Terre Valley. Soils on all of the alluvial deposits have been investigated by Johnson (1983). Even older are the relatively thin (<1 m) strath remnants and lag deposits of stream-rounded, tan-colored chert gravels on the higher, relatively flat surfaces of Paleozoic carbonate rocks that make up the Ozark Plateaus, as defined by Fenneman (1938). These deposits were abandoned by the ancestral Pomme de Terre as dissection of the Ozark Plateaus progressed. They have not been mapped in the study area and will be referred to simply as the older strath gravels, as opposed to the strath gravels of the Breshears Valley Formation of T-3. The incision of the Pomme de Terre Valley into bedrock of Paleozoic dolomites and sandstones probably occurred throughout much of Pleistocene time.

Initially I defined four Quaternary alluvial units in the lower Pomme de Terre Valley and informally called them Koch, Boney Spring, Rodgers, and Pippins alluvium (Haynes, 1978), ranging in age from late Pleistocene to Holocene. Later the deposits of T-3 were informally called the Breshears Formation (Haynes, 1978). Brakenridge (1979, 1981) redefined the Koch alluvium and renamed that part in the Breshears Valley the Trolinger Formation. The differences are discussed below.

Trolinger Spring Formation

The predominant alluvial deposit in the Breshears Valley, here named the Trolinger Spring Formation (T-2), is composed of about 9 m of gray (Munsell color 5Y 6/1 dry, 4/4 moist) silty clay and clayey silt with lesser proportions of clayey sand and gravel lenses, covering 1 to 2 m of subangular to subrounded chert gravel with a dark gray (reduced) patina. Fourteen core holes to bedrock showed Trolinger Spring Formation in the subsurface throughout the Breshears Valley (Fig. 6a). Toward the surface, approximately 14 m above the modern stream bed, the alluvium is gradational through a 10 to 20 cm mottled zone to

Figure 4. Views in the Pomme de Terre and Breshears Valleys: a. Pomme de Terre River looking upstream at Avery Bridge, a wire suspension bridge built at the turn of the century. b. Gravel pit exposure of paleosol developed on Breshears Valley Fm. on "church hill" west of Spring Creek Church. c. View to north across Breshears Valley overlooking Trolinger Spring where coring rig is in operation. d. Rodgers (T-1c) and Pippins (T-0) terraces at meander bend opposite Buzzards Bluff. e. Koch Spring during 1975 field trip of The Friends of the Pleistocene. f. Kirby Spring, 1975.

brownish yellow (10YR 6/8, 5/6) except where reducing conditions have been maintained around Trolinger, Kirby, and Jones Springs.

Throughout the Breshears Valley, the Trolinger Spring Formation typically has a well-developed dark gray soil (mollisol) (Johnson, 1983) that supported a bottomland prairie before European settlement (McMillan, 1976b). This prairie may have attracted the first homesteaders to the valley because less clearing would be necessary. The mollisol grades to a reddish-brown soil on the colluvial slopes around the periphery of the abandoned meander forming the Breshears Valley. The stratigraphic relationships revealed by trenches 76B (Fig. 5—in pocket inside back cover) indicate that part of the black color of the Pippins Cemetery Formation may be due to reworking and redeposition of the mollisol from the valley sides. Additional darkening may be due to wetting and drying in response to near surface fluctuation of the water table.

Few remnants of the Trolinger Spring Formation remain outside of the Breshears Valley along the Pomme de Terre River. Trench 77A opposite Buzzards Bluff exposed a remnant about 9 m above the streambed, which is mapped as Trolinger Spring Formation on the basis of its elevation and lithology (Brakenridge, 1979, 1981). Other occurrences (queried on Fig. 2) may exist in the mapped area, but without adequate exposure or geochronological data they cannot be positively identified. About 2 km across the river and northwest of Boney Spring there appear to be more T-2 deposits in the remnant of another cutoff meander, and yet another such occurrence is across the river southwest of Fairfield (see U.S.G.S., Fristoe 15-minute quadrangle).

Radiocarbon dating of well-preserved wood and plant fragments in Jones and Kirby Spring deposits, described later, indicates that all of the Trolinger Spring Formation is probably in excess of 32,000 years old (Table 1). At Jones Spring a finite radiocarbon date of 48,000 ± 900 B.P. (QL-962) was obtained on a juniper (*Juniperus virginiana*) root and 51,000 ± 1500 B.P. (QL-1430) on another root from peat at the base of the upper third of the formation, but these are considered minimum dates, as discussed below. An experimental uranium-series date of 160,000 B.P. was obtained on mastodon tooth enamel from the same level (McKinney, 1979) and is consistent with the assignment of a Sangamonian age to the basal peat (Haynes et al., 1983).

Pollen analyses and plant remains from peat deposits at Jones and Trolinger springs reveal an open pine parkland in the Breshears Valley during the later depositional phase of Trolinger Spring Formation (King, 1973; King and Lindsay, 1976). Fossil bones from the peat deposits include mastodon, mammoth, horse, and bison, essentially a grazing fauna (Saunders, 1977a). A basal gravelly peat at Jones Spring contains a fauna now thought by Saunders to be interglacial (personal communication).

The basal gravels in the Breshears Valley are undoubtedly channel gravels from a time when the Pomme de Terre River flowed through the valley before the cutoff of the meander occurred. Core sampling in 1973 revealed up to 50 cm of a dark gray organic clay between 8 and 9 m depth over the basal gravels in the southern part of the meander (Fig. 6a). This suggests anaerobic conditions possibly related to abandonment of the meander and formation of a slough. The overlying clays and sandy clays are vertical accretion deposits, most of which may have been deposited by overbank flooding into the abandoned meander. Gravel lenses become more frequent toward the colluvial slopes and apparently represent channels and alluvial fans from local runoff.

After abandonment of the Breshears meander and the T-2 terrace, the Pomme de Terre incised its channel about 2 m deeper into bedrock before final aggradation of Trolinger alluvium formed the next lower terrace, T-1a (Fig. 3).

The 6.7-m terrace (T-1a) at Koch Spring (Fig. 6b) was originally considered to be an erosional remnant of T-2 (Trolinger alluvium), the deposits of which were informally called Koch alluvium (Haynes, 1976). The lower elevation of the terrace containing the Koch Spring deposits, compared to the host deposits for Jones and Kirby springs, led to the suggestion that the Breshears Valley deposits might be high overbank facies of Koch alluvium (Haynes, 1978). The T-2 deposits outside of the Breshears Valley (Fig. 3, T-1a) can, however, be interpreted as being the final stage of aggradation of the Trolinger alluvium. Brakenridge (1979, 1981) has suggested a new name, Trolinger Formation, for T-2, which is formalized in the present report to Trolinger Spring Formation.

The T-1a deposits, as revealed by trenching at and adjacent to Koch Spring (Fig. 4e), consist of olive-gray (5Y 6/2, 5/2) clay and sandy clay interbedded with colluvial dolomite and angular chert gravel in the upper third of the terrace toward the dolomite hillslopes. Toward the surface the olive-gray color is mottled with yellowish-brown (10YR 5/6, 4/4) and is gradational over 10 cm to a dark, grayish-brown (10YR 4/4, 4/2), weakly prismatic soil best preserved in the footslope outcrops (Johnson, 1983). Wells drilled at Koch Spring prior to beginning our 1971 excavations encountered a meter or two of basal chert gravel over bedrock.

Most of the Koch terrace (T-1a) is covered with up to 50 cm of grayish-brown (10YR 5/2, 3/2) clayey silt that contains dispersed Archaic artifacts and is an overbank facies of the Rodgers Shelter Formation deposited on a similar facies of the Boney Spring Formation. A paleosol on the upslope part of the Koch alluvial terrace is truncated by erosional contact with the Rodgers overbank deposit, indicating that the terrace tread may have been lowered by perhaps as much as a meter by flood scour.

Radiocarbon dates for spring deposits obtained from the buried peat lens at Koch Springs (Fig. 6b) indicate an age range from about 30,000 B.P. to 32,000 B.P. (Table 1) (Haynes, et al., 1983). Pollen analyses reveal an open pine parkland during accumulation of most of the peat in the Koch Spring deposit, but a few spruce grains appear at the top of the profile (King, 1973; King and Lindsay, 1976) and are probably associated with the overlying Boney Spring Formation. There may be a degree of temporal overlap between the peat at Koch Spring and Trolinger

TABLE 1. RADIOCARBON DATES IN YEARS BEFORE PRESENT (A.D. 1950) ARRANGED STRATIGRAPHICALLY BY SITE

Formation	Radiocarbon Date (Lab. No.)		Material	Stratigraphic Position	Depth*
			Rodgers Shelter		
P_2	430±100	(A-867)	Charcoal	Late lower Pippins alluvium	
P_2	530± 70	(SMU-466)	Charcoal	Upper Stratum 4 (unit G_2)	1.13
R_6	1460± 60	(SMU-446)	Charcoal	Stratum 4, level 1	0.38
R_6	1390± 70	(SMU-438)	Charcoal	Stratum 4, level 2	0.53
R_6	1580± 70	(SMU-447)	Nuts	Stratum 4, level 2	0.53
R_5	2350± 80	(SMU-454)	Charcoal	Stratum 4, level 2	1.20
R_5	2250± 70	(SMU-478)	Charcoal	Stratum 4, level ?	1.28
R_5	2070± 70	(SMU-439)	Nuts	Stratum 4, level 3	0.69
R_5	2520± 60	(SMU-448)	Charcoal	Stratum 4, level 4	0.84
R_4	3360± 70	(SMU-510)	Charcoal	Stratum 4/3, level 6	1.14
R_4	3430± 50	(SMU-524)	Charcoal	Stratum 4/3, level ?	1.14
R_4	3530± 80	(SMU-451)	Charcoal	Stratum ?, level ?	1.43
R_3	5100±400	(M-2332)	Charcoal	Stratum 3, pit	1.37
R_3	5200±200	(M-2281)	Charcoal	Stratum 3, pit	1.37
R_2	6300±590	(ISGS-35)	?	Stratum 2, level ?	1.77
R_2	7010±160	(GAK-1171)	Charcoal	Stratum 2, level ?	3.20
R_2	8100±140	(GAK-1170)	Charcoal	Stratum 2, level ?	3.05
R_2	7490±170	(GAK-1172)	Charcoal	Stratum 2, level ?	3.44
R_2	7170±160	(SMU-502)	Charcoal	Stratum 2, level 11	1.90
R_2	7250±360	(SMU-507)	Charcoal	Stratum 2, level 12	2.06
R_1	8100±300	(A-868A)	Charcoal	Stratum 1, level ?	3.81
R_1	8030±300	(M-1900)	Charcoal	Stratum 1, level ?	4.52
R_1	10,530±650	(ISGS-481)	Charcoal	Stratum 1, level ?	8.53
R_1	10,200±330	(M-2333)	Charcoal	Stratum 1, level ?	8.98
			Rodgers Shelter Secondary Bone Carbonates		
R	105.38±2.92%M**	(SMU-533)	Secondary CO_3	Stratum 4, level 4, bone	
R_2	280±160	(SMU-530)	Secondary CO_3	Stratum 2, level 11 and 12, bone	
R_2	117.97±2.09%M	(SMU-532)	Secondary CO_3	Stratum 2, level 16, bone	
R_1	121.23±1.71%M	(SMU-526)	Secondary CO_3	Stratum 1, level 18, bone	
			Rodgers Shelter, West Terrace		
P_1	200± 60	(SMU-456)	Charcoal	Stratum 4, unit G_2	2.05
P_1	200± 70	(SMU-467)	Charcoal	Intrusive pit in Strat 4	1.98
R_6	1060±100	(SMU-474)	Charcoal	?	2.21
R_5	1910±100	(SMU-455)	Charcoal	Stratum 4, ?	2.28
R_5	2620±150	(SMU-465)	Charcoal	Stratum 4/3, ?	2.44
R_3	5130±160	(SMU-459)	Charcoal	Stratum 2, ?	3.66
R_2	7960±130	(SMU-461)	Charcoal	Stratum 1, ?	4.57
			Phillips Spring		
P_2	270± 50	(SMU-237)	Charcoal	Pit in unit H	
R_6	1410± 50	(SMU-327)	Wood	Base of unit G	1.0
R_5	1900± 80	(SMU-538)	Charcoal	Feat. 408, unit E_2	0.9
R_5	1990± 50	(SMU-234)	Charcoal	Top of unit E	0.8
R_5	2040± 60	(SMU-537)	Charcoal	Feat. 392, unit E	1.0
R_5	2250±100	(SMU-554)	Charcoal	Feat. 415, unit E	1.0
R_5	2340±180	(SMU-236)	Wood	Base of Strat. E	1.4
R_4	2910± 50	(SMU-238)	Charcoal	Top of Strat. D	1.6
R_4	3050± 60	(SMU-235)	Charcoal	On top of Strat. C_2	1.7
R_4	3330± 50	(SMU-331)	Charcoal	Feat. 3, Strat. D	1.5
R_4	3330± 50	(SMU-818)	Nuts	Deep trench, unit K_2	1.4
R_4	3650± 70	(SMU-550)	Charcoal	Feat. 424, unit K_2	1.0
R_4	3800±180	(SMU-559)	Charcoal	Sedalia comp 3, unit K_2	0.9
R_4	3750± 50	(SMU-820)	Wood	Deep trench, unit K_2	1.6
R_4	3920± 70	(SMU-558)	Charcoal	Lower Sedalia, unit K_2	1.5
R_4	3940± 70	(SMU-419)	Charcoal	Sedalia comp 1, unit K_2	1.7
R_4	3900± 60	(SMU-319)	Charcoal	Sedalia comp 1, unit K_2	1.2
R_4	3960± 70	(SMU-556)	Charcoal	Sedalia comp 1, unit K_2	1.4
R_4	4000±100	(SMU-423)	Charcoal	Sedalia comp 1, unit K_2	1.9

TABLE 1. RADIOCARBON DATES IN YEARS BEFORE PRESENT (A.D. 1950) ARRANGED STRATIGRAPHICALLY BY SITE (continued)

Formation	Radiocarbon Date (Lab. No.)		Material	Stratigraphic Position	Depth*
R_4	4240± 80	(SMU-102)	Charcoal	Sedalia comp 1, unit K_2	1.7
R_4	4220± 60	(SMU-483)	Charcoal	Sedalia comp 1, unit K_2	1.4
R_4	4310± 70	(SMU-98)	Charcoal	Sedalia comp 1, unit K_2	1.7
R_3	4980± 70	(SMU-811)	Twigs	Deep trench, unit K_1	1.4
R_3	5390± 90	(SMU-539)	Moss	Pollen Profile 74-2, unit K	1.7
R_3	5400±110	(SMU-815)	Twigs	Deep trench, unit K_1	1.7
R_2	6430± 80	(SMU-810)	Twigs	Deep trench, unit C_3	1.5
R_2	6580±180	(SMU-505)	Charcoal	Unit C_2	2.0
R_2	6900± 80	(SMU-819)	Moss	Deep trench, unit C_2	2.0
R_2	7090± 90	(SMU-557)	Wood	Unit C_2	2.0
R_2	7195±100	(SMU-812)	Twigs	Deep trench, unit C_2	
R_2	7300± 80	(SMU-726)	Wood	Monolith, unit C_2	3.3
R_2	7480± 80	(SMU-193)	Moss	Unit C_2	
R_2	7620± 90	(SMU-736)	Oak	Deep trench, unit C_2	3.0
R_2	7750± 90	(SMU-735)	Sycamore	Deep trench, unit C_2	3.4
R_2	7870± 90	(SMU-78)	Wood	1973 trench, unit C_2	3.6
R_2	7980± 70	(SMU-727)	?	Monolith, unit C_2	
R_2	8050± 90	(SMU-737)	Wood	Deep trench, unit C_1	5.4
	8350± 90	(SMU-518)	Water	Bicarbontes	
BS?	25,060±330	(SMU-813)	Humates	Monolith, unit A	5.7
	Tench 76-D (opposite shelter)				
P_2	190± 40	(SMU-500)	Leaf mat	Upper Pippins alluvium	2.4
P_2	260± 40	(SMU-432 & 508)	Wood	Upper Pippins alluvium	3.0
P_2	240± 50	(SMU-499 & 485)	Leaf mat	Upper Pippins alluvium	3.0
P_2	230± 40	(SMU-431)	Wood	Upper Pippins alluvium	3.0
R_5	1750± 90	(SMU-430)	Disp-charcoal	Rodgers alluvium	1.3
R_5	2360± 70	(SMU-506)	Disp-charcoal	Rodgers alluvium	2.0
R_4	3560± 90	(SMU-429)	Charcoal	Rodgers alluvium	2.0
	Trench 77-E (Avery Bridge)				
P_1	550±100	(UCR-822)	Charcoal	Pippins alluvium	0.8
P_1	730±100	(UCR-821)	Charcoal	Pippins alluvium	1.2
R_4	3470±130	(UCR-818)	Charcoal	Rodgers colluvium	0.8
R_4	3985±100	(UCR-819)	Charcoal	Rodgers alluvium	2.2
R_4	4585±130	(UCR-820)	Charcoal	Rodgers alluvium	2.4
	Trench 77-A (Buzzard's Bluff)				
P_1	680±100	(UCR-823)	Charcoal	Lower Pippins alluvium	1.5
P_1	820±110	(UCR-824)	Charcoal	Lower Pippins alluvium	1.5
	750±100	average of above			
	Trench 78-B (near giant hickory)				
R_4	3630± 70	(SMU-823)	Charcoal	Rodgers alluvium	0.4
R_3	5150±330	(SMU-816)	Charcoal	Rodgers alluvium	0.9
R_2	8050±100	(SMU-814)	Charcoal	Rodgers alluvium	2.7
	Boney Spring				
	Modern	(Tx-1469)	Wood	Intrusive, fallen branch	?
R_5	1920± 50	(Tx-1472)	Wood	Prehistoric pit, unit I	1.3
R_5	1910± 80	(Tx-1470)	Humates	Outside of pit, unit I	1.1
R_5	1900± 80	(Tx-1471)	Wood	Outside of pit, unit I	1.3
R_4	4200± 140	(A-1076)	Nuts	Base of unit I	1.6
R_2	7290±1900	(Tx-1466)	Charcoal	Dispersal in unit F	2.6
BS	13,550± 400	(A-1079)	Peat	Tusk filling, unit E_2	3.1
BS	13,700± 600	(M-2211)	Peat	Tuck filling, unit E_2	3.1
BS	16,190± 400	(Tx-1629)	Moss	Unit E_1 tufa	4.0
BS	28,230± 940	(Tx-1473)	$CaCO_3$	Unit E_1 tufa	4.0

TABLE 1. RADIOCARBON DATES IN YEARS BEFORE PRESENT (A.D. 1950) ARRANGED STRATIGRAPHICALLY BY SITE (continued)

Formation	Radiocarbon Date (Lab. No.)		Material	Stratigraphic Position	Depth*
BS	16,450± 200	(I-4236)	Spruce	Unit E_2/E_1	3.6
BS	16,490± 290	(Tx-1477)	Spruce	Unit E_2/E_1	3.6
BS	16,580± 220	I-3922)	Spruce	Basal unit E_2	3+
BS	16,540± 170	(Tx-1478)	Spruce	Basal unit E_2	3+
BS	20,300± 470	(Tx-1474)	Wood	Upper unit C_4	3.9
	17,320±1810	(Tx-1475)	Humates	(same sample)	
	19,550±1080	(Tx-1476)	Humates	(same sample)	
BS	21,380± 500	(Tx-1410)	Wood	Lower unit C_4	4.2
BS	20,710± 530	(Tx-1479)	Wood	Lower unit C_4	4.9
BS	22,730± 590	(Tx-1408)	Peat	Upper unit C_3	4.8
BS	28,330±3140	(Tx-1407)	Humates	Upper unit C_3	4.8
BS	26,440±1170	(Tx-1409)	Wood	Lower unit C_3	5.1
BS?	24,460±10000	(Tx-1467)	Peat	Basal unit B_1	6.2
	27,480±1950	(Tx-1468)	Humates	(same sample)	
			Trolinger Spring		
BS	20,500± 450	(I-3535)	Peat	Unit e	1.7
	17,250± 600	(I-3536)	Humates	(same sample)	
T	29,340± 900	(A-1000)	Peat	Unit d_3	1.5
T	14,450± 550	(Gx-1318)	Peat	Unit d_3	1.8
	4290± 230	(Gx-1319)	Humates	(same sample)	
T	33,220±1900-1600	(I-3599)	Peat	Unit C	2.1
BS?	21,940± 430	(SMU-170)	Collagen	Mastodon, unit d_3	
	20,100± 480	(SMU-147)	Apatite	(same sample)	
	22,330± 700	(SMU-148)	Apatite	(same sample)	
	420± 120	(SMU-145)	Secondary CO_2	(same sample)	
T	34,300±1200	(A-1080)	Peat	Unit C (tusk filling)	
BS	25,650± 700	(I-3537)	Straw	Unit b	2.3
T	38,020±2850	(SMU-935)	Peat	Unit e (?)	1.5
T	38,900±3750	(SMU-934)	Peat	Unit e (?)	1.6
T	38,280±3430	(SMU-933)	Peat	Unit d_{3b}	1.8
T	33,000±1040	(SMU-932)	Peat	Unit d_{3a}	1.6
T	32,330± 920	(SMU-931)	Peat	Unit d_{3a}	
T	>55,000	(QL-1428)	Wood	Conduit gravel	6.0
			Koch Spring		
	143.9±5.7%	(Tx-1456)	Wood	Intrusive, fallen branch	1.7
P_1	620± 80	(Tx-1411)	Wood	Pippins terrace	1.1
P_1	640± 60	(Tx-1453)	Wood	Pippins terrace	1.9
P_1	840± 60	(Tx-1454)	Wood	Pippins terrace	2.3
T	30,530± 200	(QL-1427)	Wood	Clayey peat, unit b_2	3.9
T	31,880±1340	(Tx-1412)	Straw	Clayey peat, unit b_2	4.1
T	30,880±1320	(Tx-1455)	Humates	Clayey peat, unit b_2	4.5
	31,090± 150	(QL-1429)	Wood	Clayey peat, unit b_2	4.3
T	>38,000	(Tx-1457)	Wood	Brown clay, unit b_1	5.2
T	33,550±3210	(Tx-1458)	Humates	(same sample)	
			Kirby Spring		
T	>37,000	(Gx-2718)	Humates	Top of peat	1.2
T	>57,000	(GL-1426)	Wood	Middle of peat	1.3
T	>27,000	(Gx-2719)	Humates	Bottom of peat	2.7
T	>25,000	(Gx-2720)	Humates	Bottom of peat	2.7
			Jones Spring		
T	31,130±1550	(SMU-99)	Humates	Tusk filling	1.7
T	>35,320	(SMU-94)	Peat	Upper unit e_2	1.8
	44,040±2660	(SMU-96)	Humates	(same sample)	
T	>40,000	(Tx-1627)	Peat	Top of lower peat (c_2)	3.2
T	39,020±2600	(Tx-1626)	Humates	Base of lower peat (c_1)	3.6
T	>40,000	(Tx-1628)	Wood	Olive gray clay	3.8
T	48,900± 900	(QL-962)	Root	Olive gray clay	3.5
T	51,000±1500	(QL-1430)	Root		

Values in this table, the latest laboratory corrections, take precedence over any differences with values in the figures.
*Meters below terrace tread except for Rodgers Shelter and West Terrace where it is depth below datum.
**Percent modern radiocarbon activity.

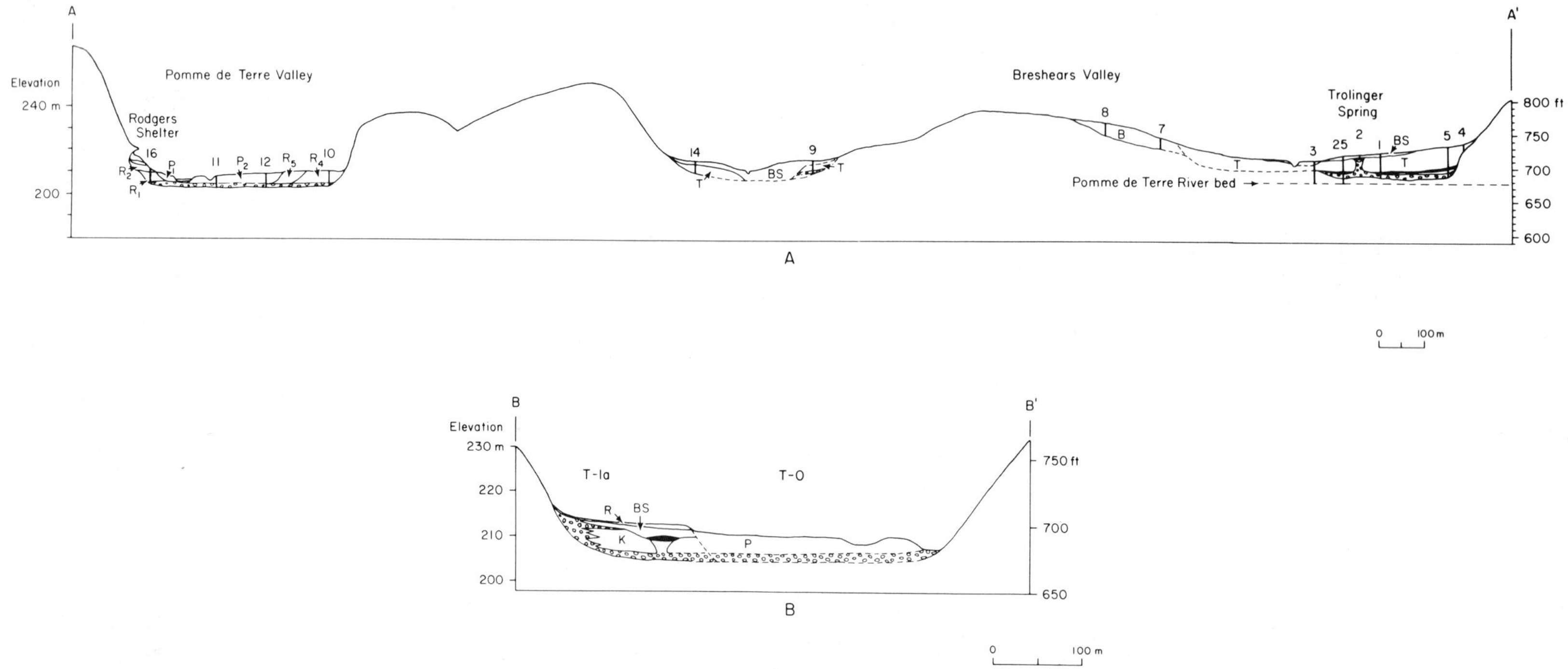

Figure 6. Geologic cross section (A) of the Pomme de Terre and Breshears Valleys constructed from core holes (numbers) and (B) of the terraces (T-1a and T-0) at Koch Spring. (See Figure 2 for location of sections.)

Spring. This will be evaluated later in the section on Trolinger Spring.

Boney Spring Formation

An erosional hiatus of 2,000 to 3,000 years separates the end of peat formation of Koch Spring from the beginning of the Boney Spring terrace (T-1b) aggradation dated 26,440 ± 1170 B.P. (Tx-1409) on wood near the base of the Boney Spring Formation at Boney Spring, 3.2 km (2 mi) northwest of the Breshears Valley (Fig. 8b). At Boney Spring (Fig. 7e) the alluvial deposits consist of 5 to 7 m of interbedded gray (5Y 7/1, 5/1) to olive (5Y 5/3, 4/1) organic clays, clayey silts, and peat over 1 to 3 m of chert pebble gravel. They are intruded by spring-laid granular tufa and clay containing spruce logs and bones of mastodon, giant beaver, ground sloth and smaller animals (Saunders, 1977a).

Boney Spring alluvium was encountered in the T-1 terrace complex 5 to 6 m above stream bed in two trenches upstream from Boney Spring. In one, trench 77B, it is overlain by a colluvial facies (Brakenridge, 1979, 1981); in the other, trench 76C (Breshears Valley), Boney Spring alluvium is interbedded with colluvial gravel and slope wash against the dolomite hillsides (Fig. 5). Exposures along Spring Branch and at Trolinger Spring reveal an erosional contact with the Trolinger Formation, and a well-developed dark gray soil (Johnson, 1983) at the top of the Boney Spring Formation around Trolinger Spring is probably a continuation of that on the Trolinger Formation.

The pollen evidence indicates initial aggradation under open pine parkland conditions changing to a spruce-dominated forest around 20,000 B.P., which lasted until 16,500 B.P. (King, 1973). The bone deposit formed after 16,500 B.P. and before 13,500 B.P. (Saunders, 1977a). At Boney Spring a hiatus of unknown duration occurs between the top of the clay (unit D) containing the spruce-dominated pollen record and the clay matrix (unit E_2) of the bone bed (Fig. 9), but remnants of an organic clay preserved in the nerve canal of mastodon tusks in unit E revealed that a mixed forest of spruce and deciduous trees existed at the spring 13,500 B.P. A hiatus therefore occurs between this date and 16,500 B.P. and indicates a period of non-deposition and most likely a significant reduction in spring discharge. The implied drought may have caused the death of several Pleistocene mammals (Saunders, 1977a).

A hiatus at the top of the Boney Spring sequence is indicated by a date of 7290 ± 900 B.P. at the base of the overlying Rodgers Formation, an overbank facies (Fig. 8b). The base of the Rodgers Shelter Formation (T-1c) at Rodgers Shelter is somewhat older than 10,500 B.P. Therefore, abandonment of Boney Spring terrace and extinction of the Pleistocene megafauna occurred between ca. 13,000 and 11,000 B.P.

Rodgers Shelter Formation

Excavation of Rodgers Shelter from 1963 to 1968 (McMillan, 1971, 1976c) and again in 1974 and 1976 (Kay, 1980) resulted in exposure of over 9 m of alluvium and colluvium forming the younger part of the intermediate terrace (T-1c) (Fig. 10b). Four strata were defined on the basis of erosional and/or lithologic contacts (Ahler, 1976) with the lowest (Stratum 1) containing Paleoindian and Early Archaic artifacts dating between 10,500 and 8,000 B.P. in a matrix of strong brown (7.5YR 5/6, 4/4) to dark grayish-brown (10YR 4/2, 2/2) alluvial clayey silt and colluvial gravel. Stratum 2 contains Middle Archaic artifacts and charcoal samples radiocarbon dated between 8,000 and 6,300 B.P. in a matrix of dark grayish-brown (10YR 4/2, 2/2) clayey silt. An erosional contact at its base truncates a dark brown (7/5YR 4/4, 3/2) soil at the top of Stratum 1. A dark grayish-brown (10Y 3/2, 2/2) soil at its top is conformably overlain by colluvial, mostly chert, gravel of Stratum 3 dating between 6,300 and 5,200 B.P. Stratum 4 is a mixture of alluvial silt and colluvial gravel, contains Late Archaic to Woodland artifacts, is divided into four subunits, and is dated at 3,600 to 1,500 B.P.

Environments of deposition and climatic inferences have been made from quantitative studies of the shelter and "west terrace" strata by Ahler (1973, 1976) and Kay (1980) and will be discussed in conjunction with data from other sections of the Rodgers Shelter Formation.

Subsequent mapping and trenching of the Rodgers Shelter Formation in the lower Pomme de Terre Valley has defined six episodes (R_1 through R_6) of cutting and filling (Fig. 3) that closely correspond to the depositional sequence in the shelter, except that a meter or so of chert gravel bedload occurs at the base of core holes carried to bedrock (Haynes, 1976, 1977). The earliest alluvium (T-1c_1) is best represented at the shelter (Fig. 10b) and the extension of the terrace to the west (Kay, 1980) by Stratum 1 which is subdivided by Ahler (1973) into 6 subunits. At the base is nearly 2 m of gray clayey, sandy silt (A^1 of Ahler, 1973) with a few dispersed clasts of stream-rounded chert gravel over bedrock dolomite. Northward toward the bluff this fine-grained alluvial sediment interfingers with a colluvial facies (A^2) of pebble- to large boulder-size fragments of dolomite obviously derived by mass wasting of the adjacent bluff. One meter of brown clayey silt (B^2) conformably overlies unit A^1 with an abrupt but gradational contact that is essentially a color change between reducing conditions in zones of saturation below and oxidation above.

Toward the bluff, unit B^2 interfingers with colluvium (B^1), that is the same as below but somewhat finer, with numerous tabular fragments of dolomite arranged parallel to the buried talus slope. This is conformably overlain by 2.4 m of brown clayey silt of unit C which is essentially a continuation of alluvial aggradation begun with unit A, except that it was deposited over talus of unit B^1 and contains thin lenses of subrounded to angular chert. Unit D, up to 1.8 m thick, is likewise a continuation of unit C aggradation but with clear evidence of pedogenesis in the form of darker color, moderate soil structure, root molds, and clay skins in a zone up to 30 cm thick. Much of this paleosol is

Figure 7. Views at springs: a. Jones Spring in 1966 showing unusual dried condition of the pond. b. Jones Spring excavations of 1975, showing white feeder sand surrounded by conduit gravels. Wall of excavations shows Trolinger Spring Fm. with well-developed mollic soil and oxidized gleying. c. Lower peat lens at Jones Spring, 1973. d. Trolinger Spring excavations of 1973 showing white feeder sand and gray conduit sand and gravel. e. View to northwest across floodplain of Boney Spring (by isolated tree near center) from the Breshears terrace (T-3). f. Boney Spring excavations of 1968 showing ring of mastodon bones around central feeder.

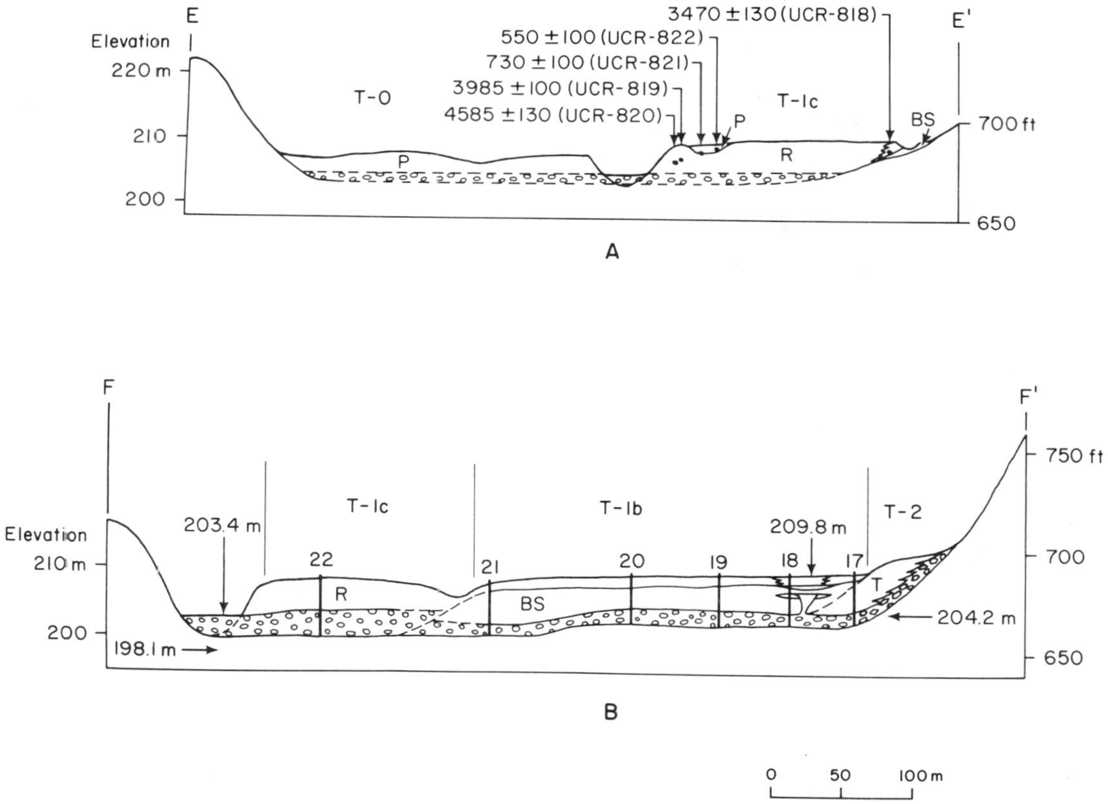

Figure 8. Geologic cross sections of the Pomme de Terre Valley (A) west of Avery Bridge showing radiocarbon dates and (B) at Boney Spring showing core holes (numbered) across the floodplain.

truncated by the contact with overlying Stratum 2, which itself is up to 1.4 m of dark grayish-brown clayey silt with dispersed pebbles, cobbles, and a few boulders of colluvial dolomite that are less tabular and unoriented as compared to the colluvium of unit B^1. Pedogenic development in Stratum 2 is stronger than that in unit D.

Radiocarbon dates from Stratum 1 (Table 1) indicate that approximately 4.6 m of alluvial silt were deposited between 10,350 and 8,100 B.P. or 0.20 cm/yr (Table 2). Then 1.9 m of Stratum 2 were laid down between 8,000 and 6,300 B.P. at a rate of 0.11 cm/yr. The radiocarbon dates leave no time for the hiatus clearly indicated by soil development followed by erosion between Stratum 2 and the top of Stratum 1. During this hiatus, the river degraded to within at least 2.8 m of the present streambed. This is indicated by a radio carbon date of 8,050 ± 90 B.P. (SMU-737) on wood from a buried channel 5 m below the surface of the Rodgers terrace (T-1c) at Phillips Spring (Figs. 10a and 11), an archaeological site on the river about 4 km southeast of the shelter. Subsequent work on the buried archaeological horizons there (Chomko, 1976; Kay, 1980) reveal that this, the second aggradational stage of Rodgers alluvium, consists of clayey silt and clay which correlates with Stratum 2 at the shelter where the youngest date is 6,300 ± 590 (ISGS-35) at the top. At both Phillips Spring and Rodgers Shelter radiocarbon dates bracket an erosional hiatus between this 6,300 date at the shelter and a date of 5,370 ± 120 (SMU-815) from the middle of unit J at Phillips (Fig. 11). Considering the shelter stratigraphy, the hiatus corresponds to the interval between the top of Stratum 2 and the base of Stratum 3. Dates from prehistoric pits dug into the alluvial fan gravels of Stratum 3 and from unit J at Phillips Spring indicate that a short episode of aggradation occurred between 5,400 and 5,000 years ago. Sometime between 6,000 and 5,400 years ago the river had degraded to at least 1.7 m below the top of the Rodgers Terrace (T-1c). It is not known if it cut all the way to bedrock.

In 1976 a backhoe trench (76-D) was excavated through the terraces across the river from Rodgers Shelter in order to have a more complete picture of the alluvial sequence. This revealed two late cut-and-fill stages of Rodgers Shelter Formation (Fig. 10b, R_4 and R_5) that consist of light brown (10YR 5/3, 3/3) clayey silt much like Stratum 1. Charcoal 2 m below the tops of each alluvial fill dated 3,560 ± 90 B.P. (SMU-429) and 2,390 ± 70 B.P. (SMU-430). The erosional contact between them was so faint that it was recognized only after proper scraping of the trench wall (Fig. 10b), but it correlates with a stratigraphic break in the lower part of Stratum 4 (unit G_1) at the shelter (Kay, 1980, pp. 82, 99–104).

Between about 0.5 and 2 m depth at the Phillips Spring site, 21 radiocarbon dates were obtained within the 1.5-m vertical interval (Fig. 11) and range from 4,310 ± 70 B.P. (SMU-98) to

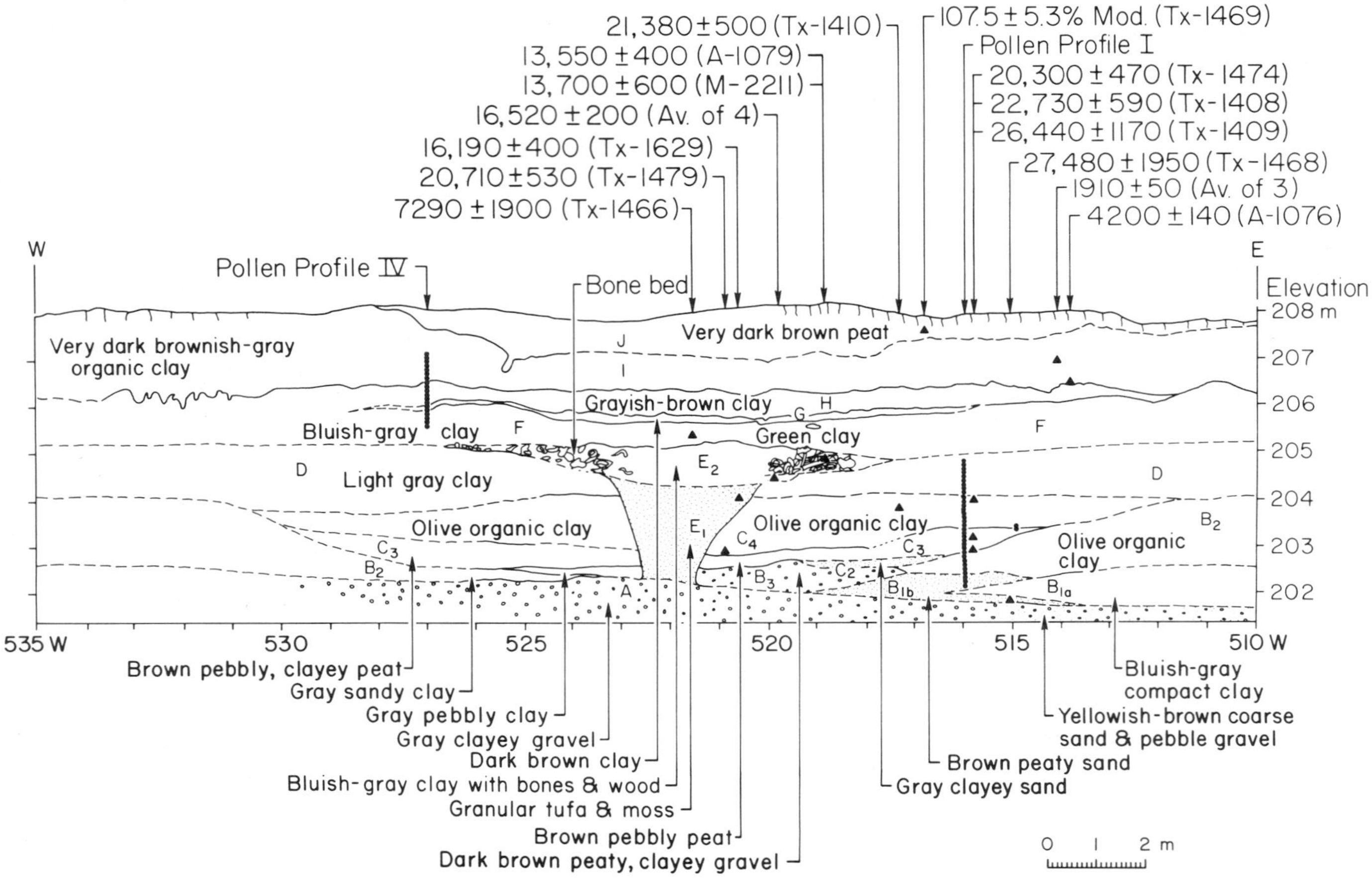

Figure 9. East-west geologic cross section of Boney Spring along lines 478N and 780N showing stratigraphic units designated by letters, position of radiocarbon dates indicated by base of black triangles, and pollen profiles indicated by lines of small circles. Units B_{1a} and B_{1b} are Trolinger Spring Fm., B_2, B_3, C_2, C_3, C_4, D, and E are Boney Spring Fm., and F, G, H, and I are Rodgers Shelter Fm. Unit J is a local deposit.

1,900 ± 80 B.P. (SMU-538). Most, but not all, of the dates are consistent with depth, but microstratigraphic relationships have been difficult to correlate from one season's excavation to another (Chomko, 1978; Haynes, 1978; Kay, 1979) because the most intensive human occupation and, therefore, bioturbation occurred during this interval between 5,000 and 1,900 years ago (Kay, 1980). The cultural disturbance is such that it would be surprising not to have some anomalies in the sequence. The situation also exemplifies the problems of trying to sort out stratigraphy with radiocarbon dates instead of the reverse, but in this case we have no choice.

A deep backhoe trench (Fig. 12) made in 1978 substantiated the stratigraphic model proposed after the previous season (Haynes, 1977) and confirmed the regional occurrence of several stratigraphic breaks seen at Rodgers Shelter and in several stratigraphic trenches. The maximum deposition rate for the Rodgers Shelter Formation of 0.63 cm/yr occurred between 8,050 and 7,750 B.P. in the lower part of the channel fill exposed by the deep trench. The lowest rate, 0.05 cm/yr, occurred in Rodgers Shelter, Stratum 4, which has a significant colluvium component. The mean rate is 0.19 cm/yr (Table 2).

Not clarified is the maximum depth of downcutting by the river during the erosional intervals following deposition of unit R_2 about 6,000 years ago (Figs. 10b and 12), but some of the erosional breaks seen in the archaeological stratigraphy are probably local scour events produced by overbank floods. Deeper channel facies, therefore, need not be expected for every break. In fact it is entirely possible that after 6,300 B.P. the river did not cut to the depths of the basal gravels during the rest of the time of deposition of the Rodgers terrace (T-1c) because no younger dates have been found more than 2.5 m below the top of the Rodgers terrace. On the other hand, trenches did not reveal the lower limits of these younger Rodgers fills (R_3-R_6), and considering the measured rates of deposition for the older Rodgers fills (R_1 and R_2) it is quite possible that the erosional hiatuses are on the order of 100 years or so.

A stratigraphic backhoe trench (77-E) (Fig. 8a) downstream from Rodgers Shelter about 1.2 km (west of the site of Avery

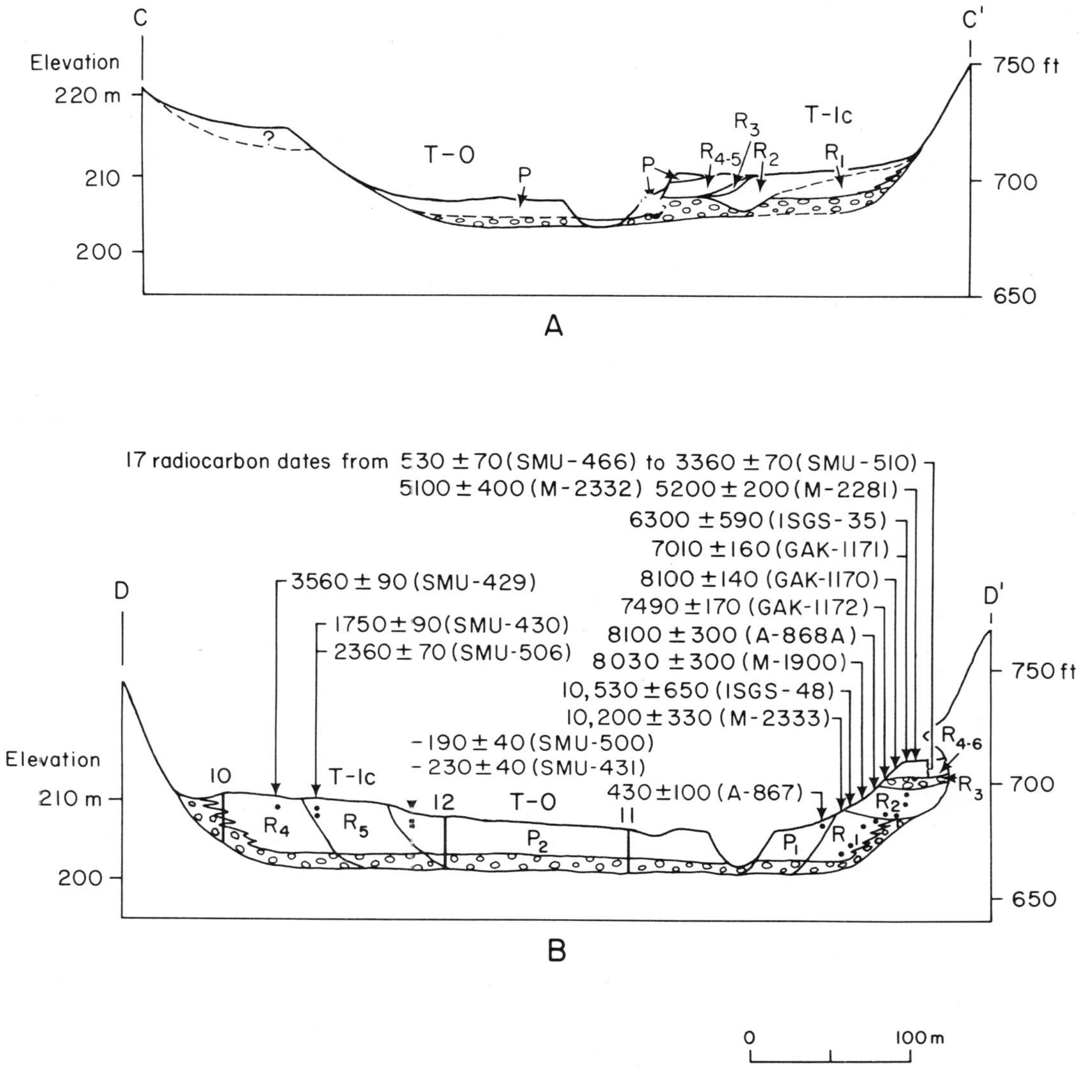

Figure 10. Geologic cross sections of the Pomme de Terre Valley (A) at Phillips Spring and (B) at Rodgers Shelter showing radiocarbon dates.

TABLE 2. ALLUVIAL DATA, POMME DE TERRE VALLEY

Formation	Average Thickness (meters)	Floodplain Cross Section Width	Floodplain Cross Section Area (m^2)*	Deposition Rate (mm/yr) Max.	Min.	Av.		Average Channel Width
Pippins Cemetery	4.0	198	548	10.0	2.0	5.0	(3)**	61
Rodgers Shelter (R_2-R_5)	6.0	396	1920	6.0	0.5	2.0	(7)	122
Rodgers Shelter (R_1)	6.0	---	---	---	---	2.0	(1)	---
Boney Spring (C_4, D)	4.6	412	1310	0.4	0.2	0.3	(5)	127
Boney Spring peat (C_3)	0.3	---	---	---	---	0.08	(1)	---
Koch peat	1.2	---	---	---	---	0.6	(1)	---
Trolinger Spring	9.1	610	3730	---	---	0.09	(1)	200
Trolinger peat (d_3)	2.0	---	---	---	---	0.03	(1)	---

*Area determined by (floodplain width x thickness) - (channel width x thickness).
**Number of determinations based upon occurrences of two or more pertinent radiocarbon dates in a particular stratum.

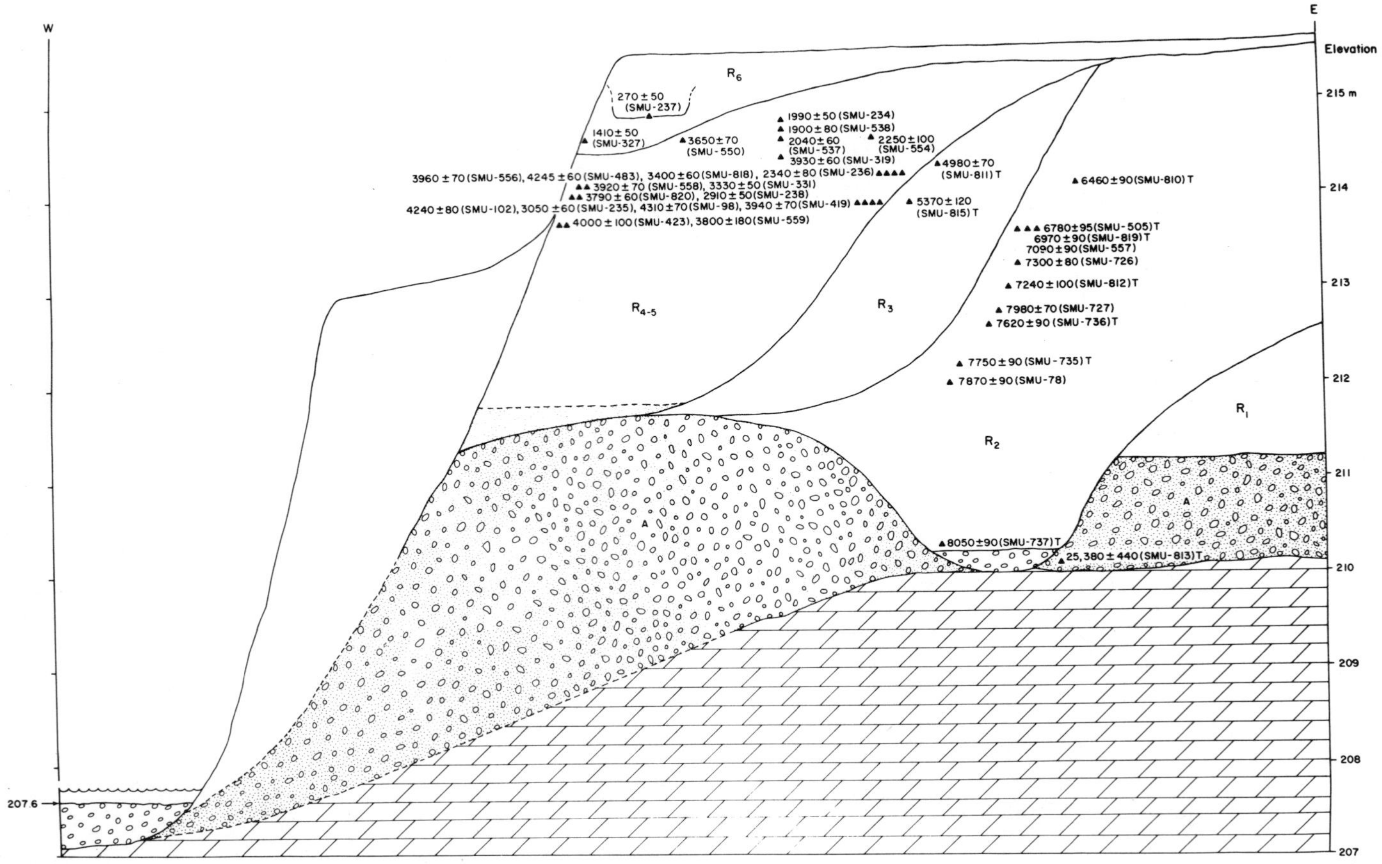

Figure 11. Generalized chronostratigraphic cross section at Phillips Spring showing the relative position of radiocarbon dates and alluvial units of the Boney Springs (BS), Rodgers Shelter (R), and Pippins Cemetery (P) formations.

Bridge) exposed several archaeological horizons between 1.5 and 2.5 below the Rodgers terrace (T-1c). Two radiocarbon dates 2.2 and 2.4 m below the surface dated 3,985 ± 100 B.P. (UCR-819) and 4,585 ± 130 B.P. (UCR-820), respectively, and were overlain by a Late Archaic projectile point (Smith type). The stratigraphic positions of the dates are compatible with the data from Phillips Spring and trench 76-D.

In trench 78-B, 300 m south of Koch Spring, stratigraphically consistent dates of 3,680 ± 100 B.P. (SMU-823) and 5,110 ± 280 B.P. (SMU-816) occur above a date of 7,990 ± 120 B.P. (SMU-814). No erosional contact was observed between the latter two (Brakenridge, 1979) but, as just discussed, the evidence for an erosional episode between about 6,000 and 5,500 B.P. is clear elsewhere. Another Smith point was found slightly below the youngest date which is consistent with the one from trench 77-E.

From the cross section of the Pomme de Terre Valley in trench 76-D and at Rodgers Shelter (Fig. 10b), it is clear that the river did not aggrade to the level of Ahler's (1973) Stratum 4 which stands 1 to 2 m above the top of the Rodgers compound terrace tread unless this surface has been reduced by scour. The flatness of the surface and its uniform soil do not suggest this. It is probable that what alluvial component Stratum 4 has was derived from relatively high, and probably infrequent floods, and occasionally from slope wash resulting in a slow deposition rate of only 0.5 cm/yr (Table 2).

The only observed coarse-grained facies of the Rodgers Shelter Formation, other than the basal gravels and colluvium, is a sandy, medium pebble chert gravel exposed in trench 77-A (Fig. 4d). Brakenridge (1981, Fig. 3C) interprets this as a relict point bar. The absence of clearly discernable point bar stratigraphy (meander scrolls) may indicate the reworking of older gravels. The deposit appears to be unique but other meanders along the river were not trenched to test this possibility.

Abandonment of the Rodgers terrace (T-1c) occurred after 1,600 B.P., according to the date from trench 76-D, but may have been after 1,400 B.P. from the evidence at Phillips Spring where fine-grained alluvium of this age (R_6) occurs near the top of the sequence (Fig. 11). Next to the bluffs a black soil (Mollic Udifluvent: Johnson, 1983) occurs at the top of colluvial facies of the Rodgers Shelter Formation.

In order to test for the migration of modern carbon in soil and groundwater-deposited carbonates, the secondary carbonate in bones within the Rodgers Shelter sequence was analyzed for

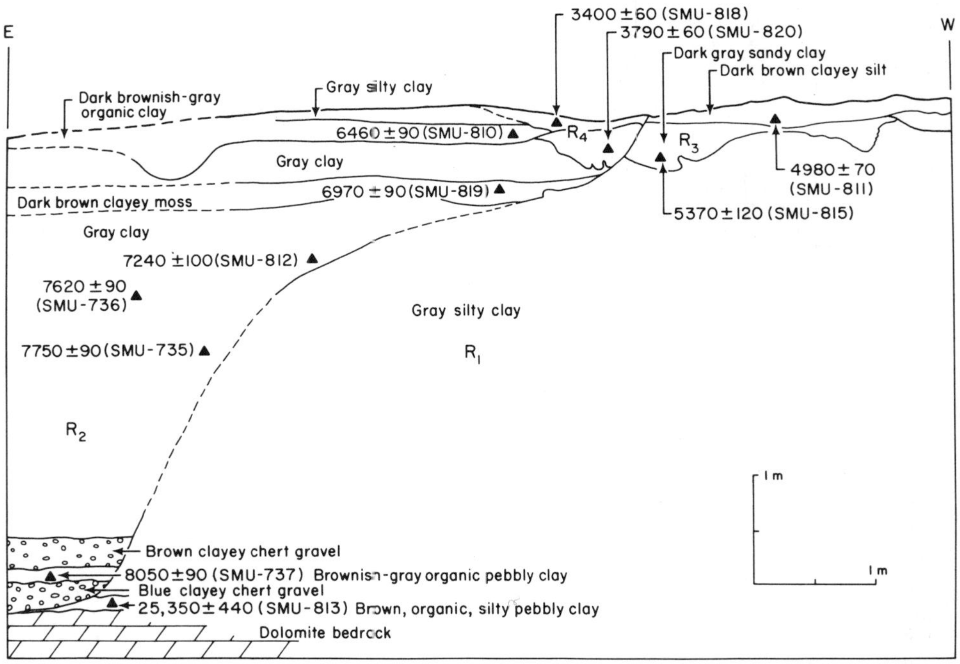

Figure 12. Stratigraphy of a deep backhoe exposure at Phillips Spring showing the relationships of 11 radiocarbon samples.

radiocarbon. The results (Table 1) show penetration of pedogenic nuclear-age radiocarbon to depths of at least 0.6 m, whereas groundwater has fluctuated to at least 6 m above its normal level since nuclear testing began in 1950. It is not known whether this represents the zone of seasonal water table fluctuation or is the result of less frequent recharge events.

*Pippins Cemetery Formation**

The historic flood plain (T-O) (Fig. 4d) of the Pomme de Terre is also a compound, 4- to 6-m terrace of two cut-and-fill deposits composed of gray silt or silty fine sand over brown to gray sand over brown chert pebble gravel. Buried slough deposits containing well-preserved wood and leaves occur in both deposits and have provided several radiocarbon dates (Table 1). The oldest radiocarbon date, 840 ± 60 B.P. (Tx-1434), is in the early Pippins Cemetery Formation (T-Oa) exposed by a backhoe trench at Koch Spring (Fig. 13, V-5), and the youngest, 430 ± 100 B.P. (A-867), is on charcoal 50 cm below the top where exposed by a bulldozer trench at Rodgers Shelter (Fig. 10b). On the basis of similar evidence from trench 76-D, the late Pippins Cemetery Formation (T-Ob) began to aggrade shortly before 300 yr B.P. and was still an active flood plain until completion of the dam at Lake Pomme de Terre in 1960 prohibited further flooding. Overbank deposition occurred on T-1c as late as 270 years ago at both Phillips Spring and Rodgers West Terrace, and local residents report all of T-1 was under several feet of water on August 8, 1927 during the highest flood of living memory. In 1838, the time the Breshears moved into the valley, T-Ob was probably much as it is today.

*The name is spelled Pippens on the Fristoe 15-minute quadrangle map (U.S.G.S., 1942), but the family spells their name Pippins.

GEOCHRONOLOGY OF SPRINGS

Five fossil bone-bearing artesian springs in the lower Pomme de Terre Valley form a rough NNW alignment (Fig. 2) and are characterized by peat lenses over sand feeders surrounded by chert gravels penetrating late Pleistocene alluvium. Our investigations from 1967 to 1979 greatly expanded knowledge of the age and anatomy of concentric-feeder or eye-type springs in general and of associated flora and fauna of the Pomme de Terre area in particular.

Koch Spring, the most southerly, produced part of the collection of mastodon bones that Albert Koch sold to the British Museum in 1844. Kirby Spring, where additional mastodon bones were uncovered during the latter half of the 19th century (McMillan, 1976a), lies in the southern arm of the Breshears Valley 2.7 km to the northwest of Koch Spring (Fig. 2). Trolinger and Jones Springs lie less than a kilometer east of Kirby, but fossil bones were not discovered in them until our test excavations of 1968 and 1971, respectively.

A spring 5 km northwest of the Trolinger is known locally as Boney Spring because of the large bones that appeared at the surface on several occasions, according to local verbal history. Fossil ivory was identified from a core sample for pollen analysis in 1965 (Mehringer and others, 1968) and bones of 31 masto-

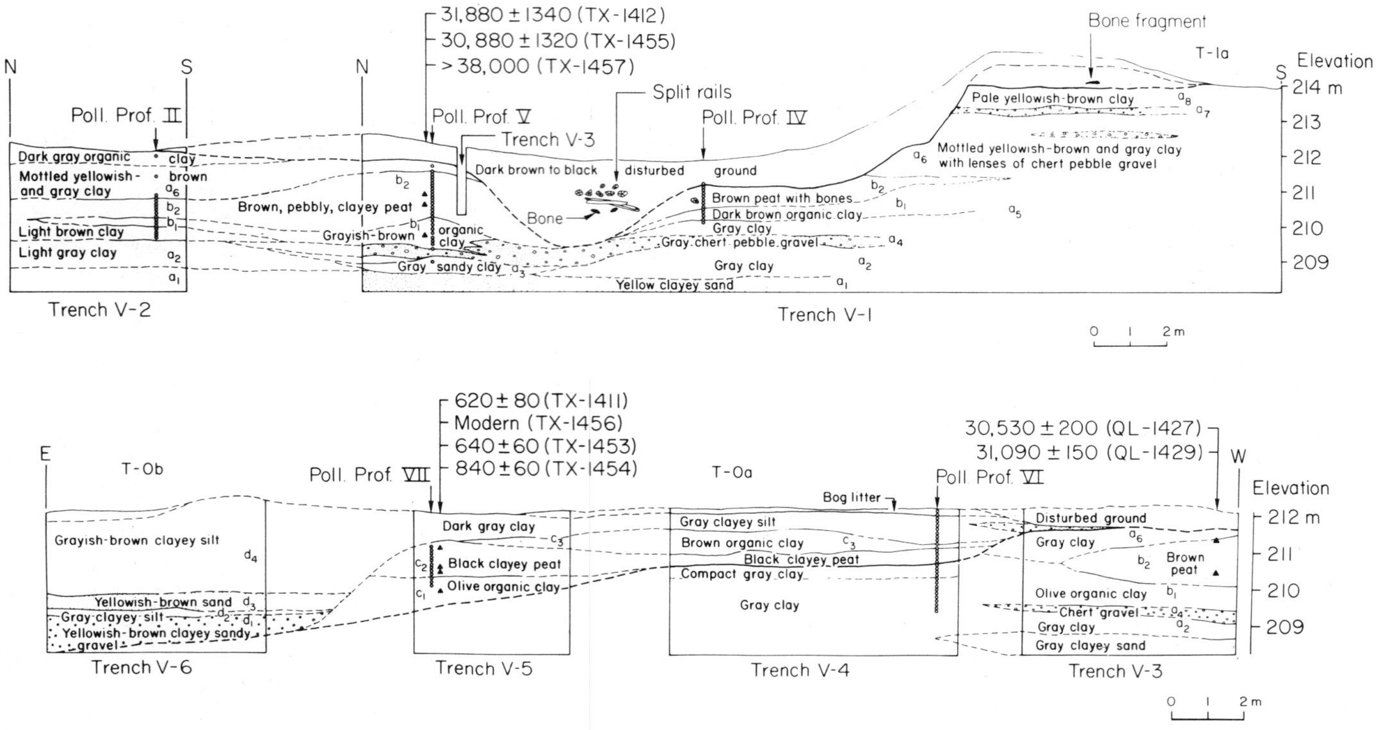

Figure 13. Stratigraphic profiles of trenches at Koch Spring showing positions of pollen profiles and radiocarbon dates.

dons were recovered in subsequent excavations (Saunders, 1977a) at Boney Springs. A more prominent spring on the west side of the Pomme de Terre River 3 km north-northwest of Boney Spring was the site of a mid-1900s resort known as Nigger Springs Camp. All of these springs lie on a northwesterly alignment, so the possibility of a related fault is real, though not proven.

Trolinger Spring

Before scientific excavations began in 1967, Trolinger was simply a wet, grass-covered, shallow depression near the east bank of Jones Branch which received a weak seepy discharge from the spring until 1964 when, according to Ivan Trolinger, the former owner, it ceased to flow two days after the Alaskan earthquake of March 27. Verbal history also recorded that at one time a tractor became stuck in the spring.

Our excavations proceeded from a north-south bulldozer trench placed along the west side of the spring that provided both drainage and an initial look at the stratigraphy (Fig. 14a). Most of the fossil bones (Fig. 14b) occurred at the base of a lens of dark brown sandy peat (Fig. 16c, unit d_3) dating between 34,330 ± 1,200 B.P. (A-1080A) and 29,340 ± 900 B.P. (A-1000). They were disconformably overlain by a dark brown clayey peat (unit e) with a date of 20,500 ± 450 B.P. (I-3535).

A mixed layer (unit c) of peat, striped with lenses of white feeder sand (Fig. 15a), occurred between the peat and underlying white sand feeder (unit b). All of these springlaid deposits are overlain by gray silty to sandy clay of the Boney Spring Formation (Fig. 16, units d_2, f_2, and g—in pocket inside back cover).

In 1978 and 1979 renewed excavations were conducted at Trolinger to recover the rest of the bones taphonomically and to clarify the stratigraphic relationship of the spring deposits to the alluvium. From the results (Fig. 16b and c) it is apparent that the compound peat lens rests on an erosional unconformity separating the Trolinger Spring Formation from the Boney Springs Formation. The sand feeder (Fig. 7d) is surrounded by sandy, angular, blue chert gravel (Fig. 15b) that has intruded, by hydraulic pressure, Trolinger alluvium either by stoping or injection along a line of weakness. On the final day of excavation deep backhoe trenches were made below the level of hand excavation which revealed the configuration of the gravel column as rising from the basal gravels underlying the Trolinger Spring Formation (Fig. 18a).

Dispersed within the gravel column were pieces of wood, fragments of cellular tufa, and small masses of moss. These could have been brought up from the dark-gray organic layer, found earlier by our coring operations to overlie the basal gravel. If they were brought up by the intruding gravel, the organic carbon should be greater than 51,000 B.P., the oldest date obtained from the upper Trolinger alluvium at Jones Spring (Table 1). A radiocarbon date of >55,000 on wood confirms this interpretation (Fig. 16b). Compatible with this are uranium series dates by Szabo (1980) of 76,000 B.P. for bone and over 100,000 B.P. for

Figure 14. Panoramic views of Trolinger Spring: a. Overview looking east. b. The bone bed in peat d_3, unit c, and unit b.

the tufa fragments in unit b (samples 7 and 10, respectively, of Fig. 16a, line 30W).

The sand-filled feeder is undoubtedly the result of artesian discharge through the gravel. The poor sorting and angularity of some of the chert gravel, the 3-m diameter of the feeder before flaring out at the top, and the 6–8-m diameter of the gravel column all suggest forceful injection of gravel fluidized by water under considerable hydrostatic pressure. Figure 17 shows a proposed sequence of spring development where gravel occupies a cylindrical conduit and winnowed sand forms a conical feeder. The process may start by water working its way from the gravel aquifer along some weakness (a). As the ascending flow erodes the clay-silt matrix and widens the conduit, gravel is swept into it (b). Interstitial sand is winnowed out of the gravel and carried farther up the column. Gravel lenses within the host alluvium and encountered by the ascending water have their fines winnowed out and remain in the conduit as lag gravel. At Trolinger Spring the gravel draped over the edge of the conduit under the northern edge of the peat (Fig. 16a, lines 30W and 32W) has been affected in this manner. This erosional process would continue until the size and configuration of the feeder reached a steady state adjusted to the discharge (c). At this stage the spring is an open pool of clear water at the bottom of which is a bed of roiling, white, well-sorted sand with dark rings of waterlogged organic matter around active centers (d). As moss, pond weeds, and other vegetation around the pool die, fragments are pushed to the periphery, become waterlogged, and sink in the area of quiet water. Thus, a raised rim forms from dead vegetation and sediment trapped therein. As discharge declines, the feeder sand compacts, and discharge occurs through a smaller subfeeder (e) and vegetation encroaches over inactive parts of the older feeder (e). Eventually a lens of peat covers the entire spring, and greatly reduced discharge works its way to the surface through a subfeeder penetrating the peat (f).

Microstratigraphy within the gravel column is made up of steeply dipping to vertical contacts separating gravels of either different shades of gray and/or grain size suggesting different episodes of movement (Fig. 7d). Similar streaks in the sand feeder represent subfeeders and indicate reduced discharge. Figure 15c shows the same phenomenon at Jones Spring. As flow velocity continues to decrease, dead vegetation and other organic matter settle to the bottom and the peat layer begins to accumulate over

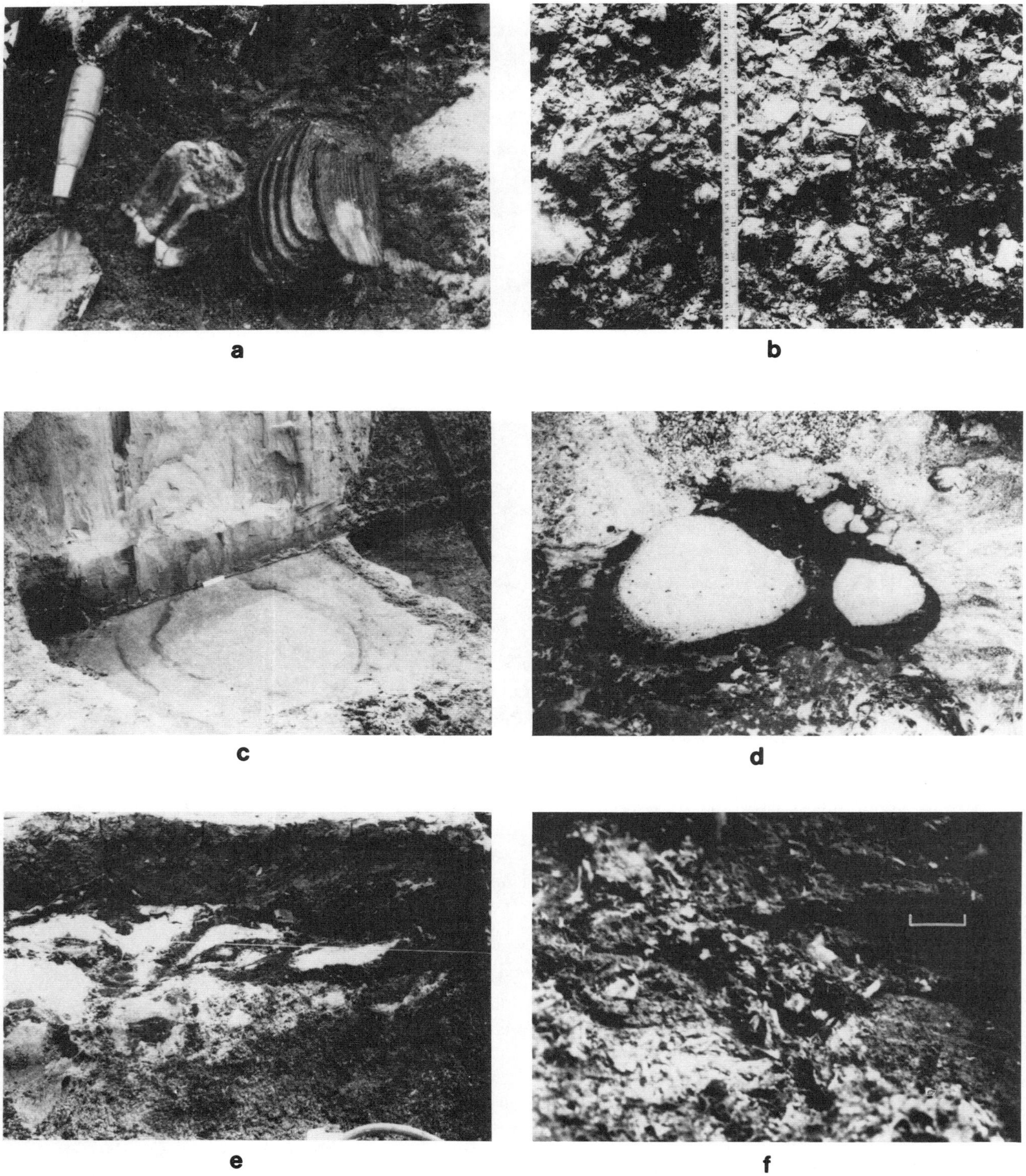

a b c d e f

Figure 15. View of spring sediments: a. Molars of both mastodon and mammoth in upper peat (unit e) at Jones Spring. b. Bluish-gray chert gravel in conduit at Trolinger Spring. c. Feeder (unit d_2) at Jones Spring showing concentric stratigraphy and subfeeders. d. Subfeeders at Jones Spring reactivated by discharge due to pump failure. Concentric black zones are cominuted peat fragments. e. Variegated sand and peats (units e_1 and e_2) at Jones Spring were also common at Trolinger Spring. f. Masticated (?) plant fragments in peat (b_2) at Koch Spring.

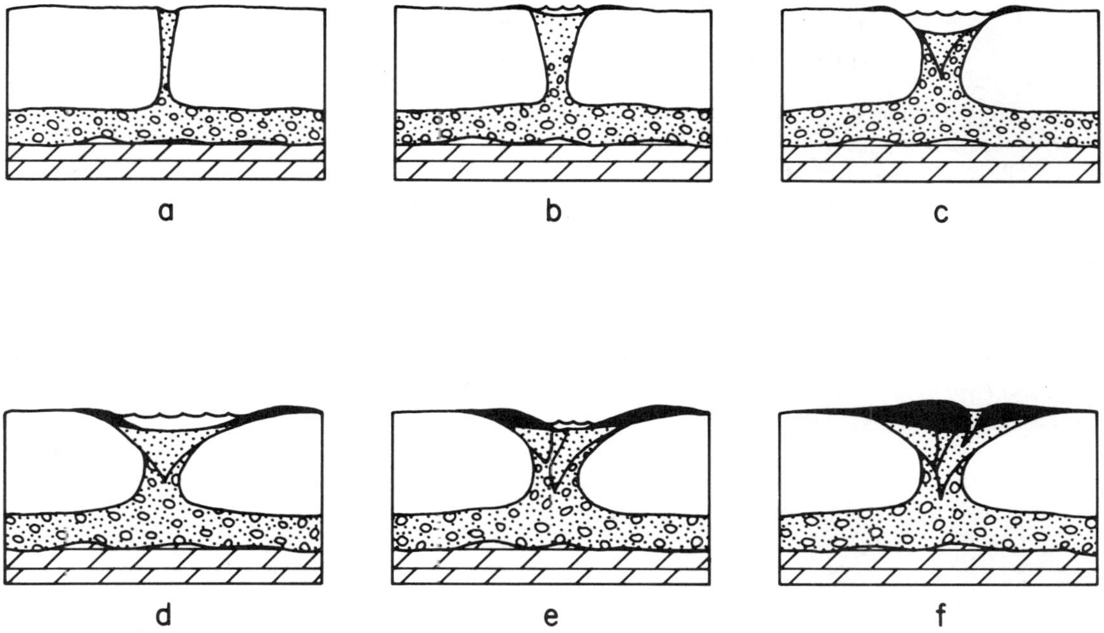

Figure 17. Diagrammatic sections showing hypothetical sequence of spring development: a. Initial ascent of water from the gravel aquifer to the surface. b. Conduit widens as sand and gravel is swept upward. c. Widening of conduit and winnowing of feeder sand. d. Formation of open pool at maximum discharge. e. Accumulation of peat, vegetation mats, and subfeeders at reduced discharge. f. Compaction of peat and greatly reduced discharge via subfeeders.

the white sand. Rings of waterlogged vegetation around active subfeeders were seen in miniature as our excavations exposed the feeder sand. This phenomenon was best displayed at Jones Spring when a pump failure caused resurgence of discharge (Fig. 15d).

Pulsating variations in discharge in subfeeders can cause some white sand to be deposited as thin irregular lenses over peat. Unit c at Trolinger is just such a deposit of "zebra striped" white sand and black (dark brown) peat (Fig. 14b). Peat (unit d_3) overlying this is streaked with a few thin layers of sand and is penetrated by a few small subfeeders (Fig. 15e). The microstratigraphy is complex, and irregular isolated patches of peat appear to be derived from an earlier peat (Fig. 16a, line 10W). Truncational contacts within d_3 (Fig. 16, lines 34 and 35W) could be small compaction faults or could represent erosional episodes, but the fact that they could not be traced for more than about 3 m is consistent with the compaction-fault hypothesis. Radiocarbon dating of peat either side of the contact between units d_{3b} and d_{3c} reveals a reversal that is inconsistent with the erosional hypothesis. Mixing has occurred but the mechanism is not obvious. Uranium series dates (Szabo, 1982, and personal communication) are 36,000 (unit c) and 33,000 B.P. (unit d_3) on the sediment (Fig. 16a, line 30W, samples 8 and 3, respectively), in good agreement with the radiocarbon dates, and 22,000 B.P. on bone (Fig. 16a, sample 3), also in good agreement with radiocarbon dates on bone samples from unit d_3 (Table 1). Another uranium series date of 15,000 B.P. from the top of unit d_3) (Fig. 16a, sample 9) is significantly younger and could be intrusive.

Another apparent inversion in the dates is shown by a radiocarbon date of 25,650 ± 700 B.P. (T-3537) on straw and twig fragments found at the top of the feeder (Fig. 16c, lines 40 and 41W, unit b) and below the older peat (Fig. 16c, line 38W, unit d_3) dating 34,300 ± 1,200 B.P. (A-1080A). I have suggested that the straw and twig fragments, whose yellow color was in contrast to the dark brown peat, may have been stomach contents of a mastodon that fell into the spring and sank through the peat 10,000 years after the peat formed (Haynes, 1976). This hypothesis is further supported by the three radiocarbon dates on bone, statistically averaging about 22,000 B.P. (Table 1) and the two uranium series dates of the same age (Fig. 16a, samples 8 and 3).

A more obvious erosional contact separates the older peat (unit d_{3c}) from unit e (Fig. 16, lines 34 and 35W), the younger peat at Trolinger. The stratigraphy suggests an episode of reduced discharge followed by a return to discharge about like that associated with the lower part. Had stronger discharge occurred after the erosional interval I would expect significant amounts of feeder sand to occur on the contact between them, but this is not the case. The radiocarbon dates show no significant age difference between unit d_{3c} and unit e (Table 1).

One can envision Trolinger Spring 26,000 to 22,000 years ago as a circular area of moss-covered peat, like the iris of the eye, over saturated feeder sand and surrounding a 'pupil' of clear water seeping away from the spring through the spongy moss and semiaquatic plants (F. King, personal communication). The vegetational mat would be weaker farther out from the conduit wall and closer to the open water. Any large animal approaching the open water would find itself literally on shaky ground and if too far out on the vegetational lip its leg would readily penetrate the mat. Struggling would further break up the supporting mat, causing the whole body to sink into the underlying quicksand. Anyone who has walked out onto such a spring will be fully aware of what an excellent natural trap it can be for large animals seeking water, the natural bait.

With further reduction of discharge the 'pupil' would become smaller in diameter or completely overgrown as the remaining discharge ascended through subfeeders. With little or no discharge the peat would compress and the feeder sand would compact, reducing the trap hazard. Such was Trolinger Spring when we first observed it in 1966.

Radiocarbon dates indicate significantly reduced discharge after 32,000 B.P. Perhaps it was during this relatively dry period that some animals fell through the peaty vegetation mat around the reduced 'eye' of Trolinger Spring.

The Farmdalian Substage, 22,000 to 27,000 B.P. of the Midcontinent and the Plum Point Interstade, 22,000 to 32,000 B.P., of the eastern Great Lakes area are within this period, which suggests a climatic cause for the reduced discharge. This is supported by the pollen record, which shows a shift from nonarboreal pollen (NAP) and pine to spruce dominance at the top of unit e (King and Lindsay, 1976). The radiocarbon dates of 20,500 $\pm$ 450 B.P. for the middle of unit e and 14,450 $\pm$ 550 B.P. for unit d_3 (Fig. 16c) may be too young because contamination by younger humates is indicated by the significantly younger humate dates for both unit e and d_3 (Table 1). Samples collected in 1978 were given more rigorous laboratory treatment, so their radiocarbon dates of 32,000 to 38,000 B.P. supports the hypothesized contamination.

A uranium series date of 18,000 B.P. on bone from near the basal contact of the gray clay, unit d_2, (Fig. 16a, sample 2) overlying the peat units is stratigraphically consistent with respect to the radiocarbon date of 20,500 $\pm$ 450 B.P. in unit e but somewhat inconsistent with respect to the bone date of 15,000 B.P. at the top of unit d_3. Once again, intrusion is invoked as the explanation if the dates are correct.

The contact between the top of the peat units and the overlying gray clay (unit d_2) is gradational over 5 to 10 cm, but as stated earlier the peat pinches out between unit d_2 and a chert gravel lens that drapes over the edge of the conduit (Fig. 16a, line 30W). The thin (0-15 cm) gravel layer rests on a sharp erosional contact with the underlying clay and gravels of the Trolinger Spring Formation. The stratigraphic situation indicates that 1 to 2 m of Trolinger alluvium was stripped along Jones Branch, and probably Spring Creek as well, before accumulation of the peat at Trolinger Spring began (Fig. 18a). The thin gravel layer may be a lag of colluvial gravel resulting from this erosion. Accumulation of the overlying clay must have begun relatively soon thereafter because there is no significant buried paleosol on the contact.

The peat accumulation appears to have been before and in part penecontemporaneous with deposition of unit d_2 gray clay, which may include a component of clay winnowed out of the spring, but its widespread extent in the Breshears Valley indicates an alluvial origin, probably the result of occasional high floods into the abandoned meander.

A late Wisconsinan (Woodfordian) age for units d_2, f_2, g, and h is supported by the radiocarbon dates from the upper peat (unit e) at Trolinger as well as the pollen analysis from unit f_2 (King and Lindsay, 1976). They represent the Boney Spring Formation in the Breshears Valley. As the peat accumulated from vegetation around the 'eye', it must have been covered from time to time by flood-laid clay. Elsewhere in the Breshears Valley interbedded lenses of silty clay and chert gravels in the Boney Spring Formation were undoubtedly washed in from adjacent slopes (Fig. 5).

The varying discharge of Trolinger Spring was undoubtedly related to climatic changes, at least in part, but the origin of the spring could have been tectonic. There is no direct evidence of a fault through the northwesterly alignment of springs in the lower Pomme de Terre Valley, but it may be significant that all of the principal springs except Phillips are more or less on this alignment and all have an unusually high $^{234}U/^{238}U$ ratio as compared to most natural waters (Szabo, 1982). In any case, glacial episodes would be times of greater hydrostatic pressure and discharge. Unfortunately, a radiocarbon date for the period of maximum discharge represented by the gravel-filled conduit at Trolinger is beyond the present range of the method (Table 1), but it could have been contemporaneous with that at Jones Spring. In this regard, an age between the 51,000 B.P. date at Jones (Fig. 19a, line 2.9S—in pocket inside back cover) and a uranium-series date of 76,000 B.P. (Szabo, 1982, and personal communication) from the conduit at Trolinger (Fig. 16a, sample 7) would be a reasonable estimate. It was during this time that discharge broke through to the surface, and if this was in response to increased recharge during a glacial substage it would have been an early Wisconsinan substage. On the other hand, if the lowest part (unit c_1) at Jones is Sangamonian, the increased discharge may have been during an Illinoian substage.

This leaves a substantial hiatus between the conduit gravels and the overlying peat and raises a question of whether there had been an earlier peat deposit at Trolinger Spring, perhaps contemporaneous with the lower peat at Jones Spring. The fragments of peat within the older peat at Trolinger lend credence to this possibility, which will be considered again when Jones Spring is discussed.

The erosional interval between the Trolinger Spring and Boney Spring formations occurred probably as a result of lowering of local base level at the entrance to the Breshears Valley sometime before 38,000 B.P. and may have caused the destruction of the first peat at Trolinger. Presumably this was a time of reduced discharge of the Pomme de Terre that occurred within middle Wisconsinan time and perhaps correlates to the Port Talbot Interstade (Terasmae and Dreimanis, 1976). Later, when the cycle swung again toward increased discharge, hydrostatic pressure on the basal gravels of the cutoff meander increased to the

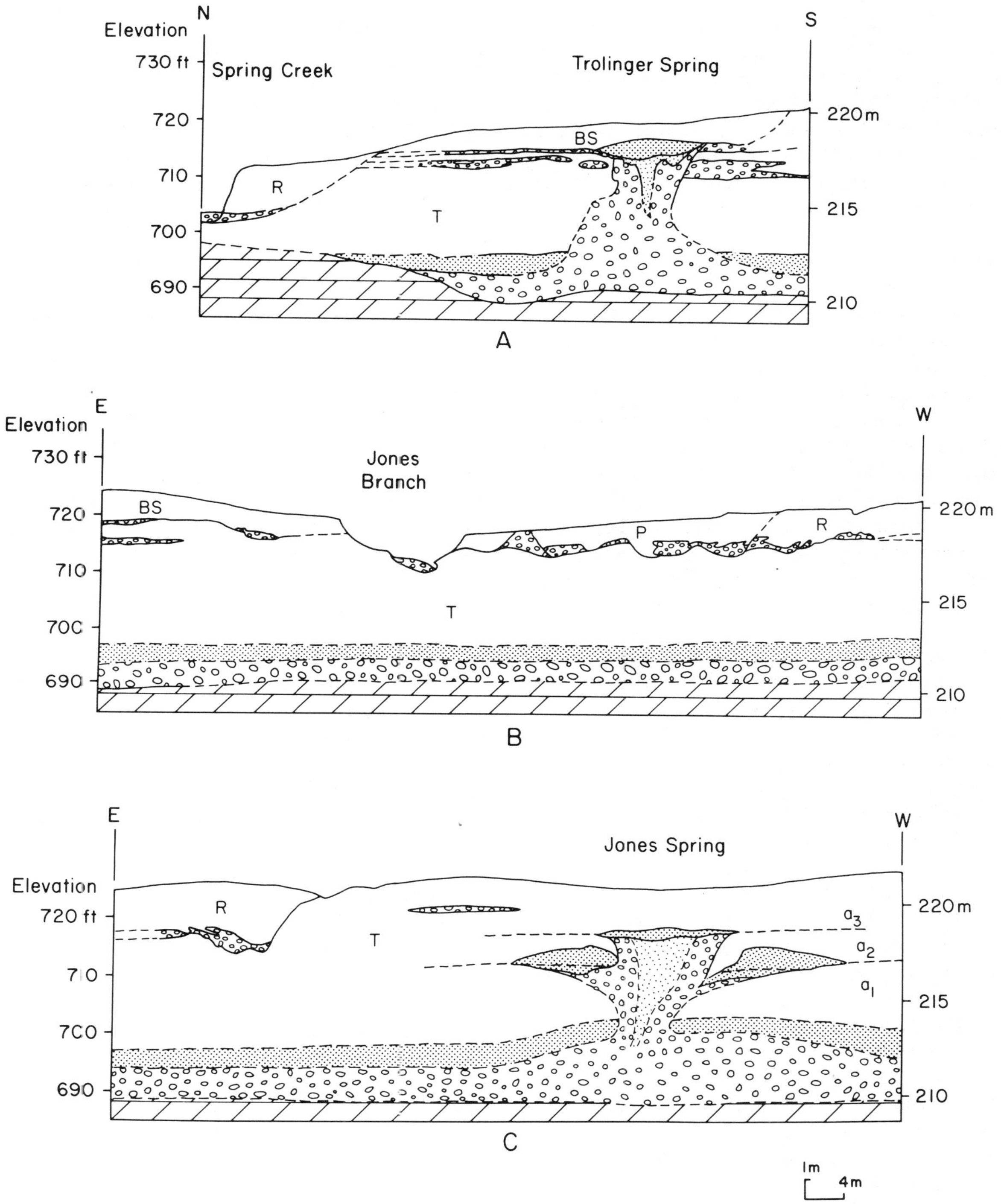

Figure 18. Geologic cross sections: (A) from Spring Creek to core hole 1 through Trolinger Spring showing organic slough deposits at the base of the fine-grained facies of Trolinger Spring Fm., (B) Trench 79B showing alluvial deposits along Jones Branch, and (C) through Jones Spring showing upper a_3, middle a_2, and lower a_1 Trolinger Spring Fm. and basal organic layer. East end of C adjoins west end of B. (See Fig. 2, inset). Slough and peat deposits stippled.

point where Trolinger Spring may have been rejuvenated between 32,000 and 38,000 B.P. The older peat might correlate to the Port Talbot I Interstade and the younger peat might be equivalent to Port Talbot II. Or, as stated previously, the younger part of the peat might correlate with the Plum Point Interstade and the older with the entire Port Talbot, in which case the inferred rejuvenation of spring discharge would correlate with the Cherrytree Stade.

Jones Spring

When first examined in the dry summer of 1968, Jones Spring was simply a dried-up pond (Fig. 7a) across Jones Branch, and about 120 m southwest of Trolinger Spring. Discharge into the branch was via a small gully, a meter or so below the rim of the pond on the northeast side. Francis Kirby, former owner, said the pond had been made by deepening a wet spot in his field with a bulldozer (J. E. King, personal communication). In the summer of 1971, when the pond had its more normal complement of water, we drained it through a backhoe trench on the northeast side and dug trenches from near the center to beyond the west side, revealing nearly a meter of brown peat over chert gravel and under gray and yellowish-brown clay of the Trolinger Spring Formation.

The presence in the Jones Spring deposits of both mastodon and mammoth bones was established in 1973 when full-scale excavations began. Two peat lenses were discovered, one above the other, with indications that the larger and lower peat deposit (Fig. 7c) contained the mammoth bones, whereas the mastodon bones were confined to the upper peat. However, subsequent excavations (Fig. 7b), 1974 to 1977, provided excellent stratigraphic profiles (Fig. 19), revealing elements of both animals in both peat lenses. The sequence of events leading to the two deposits can be read from the stratigraphy.

The lower peat, a 15-m diameter lens of brown, fine-grained peat over 1 m thick (unit c), overlies a 4- by 7-meter diameter conduit (Fig. 19a) filled with roughly concentric strata of sandy chert gravel (unit b) much like the configuration at Trolinger Spring. Bones of mastodon, mammoth, and bison are clearly associated with the lower peat (Saunders, in preparation), and a weak gradational contact separating the upper two-thirds (unit c_2) from a gravelly lower third (unit c_1) may also separate two different species of bison according to Saunders (personal communication). The upper peat (unit e) (Fig. 15e) also overlies conduit gravels (unit d_1) but these are clearly redeposited, at least in part, from the earlier conduit because they pass completely through the lower peat (Fig. 19b). The forceful nature of this intrusion is indicated by micro-reverse faults observed in the lower peat on the west side (Fig. 19b, line 0.0 North). The inner conduit gravels display concentric strata but with considerable convolutions. The inner core consists of white, well-sorted medium quartz sand (unit d_3), a typical sand feeder with accompanying subfeeders well displayed by the microstratigraphy (Fig. 15c).

The mixed nature of the conduit sediments is manifested by dispersed fragments of peat, wood, and bone. Some of the wood and bone fragments show considerable abrasion and may have been swept up with the gravels. Even the feeder, despite the high degree of sorting displayed by the sand, contained an isolated mastodon tooth, a large tusk fragment, and half of a mastodon jaw complete with teeth. These items apparently sank into the feeder, which would have been quicksand when still active. That they were not carried up in the feeder is apparent from the contrast in grain size between the sand and the foreign objects. The sand is sorted by the ascending water, and the grain size is adjusted to the average setting velocity following Stokes' law.

The upper peat (Fig. 15e) is also a mixture, but of a very dark brown (10YR 2.5/1) to black, very organic clay (unit e_2) riddled with irregular patches of white feeder sand (unit e_3, not differentiated on Fig. 19) in places, over a striped mixture of sand and peat (Fig. 19a, unit e_1) some of which must be reworked peat, sand, and clay from below. Abraded fragments of wood, bone, and teeth may have been brought up from below during the early stages of ascent of the inner conduit gravels, but the relatively fresh, well-preserved condition of some of the fossil bones suggests contemporaneity with the upper peat. Both mastodon and mammoth bones are a part of this association (Fig. 15a).

A dark gray clay (unit e_4) overlying the organic clay (unit e_2) is laminated in places and apparently represents bottom sediments of a spring-fed pond when discharge must have been slight or perhaps nil (Fig. 19b). I did not observe the upper contact of this unit because of its removal by mechanical equipment before I was on hand, but from other witnesses' accounts it was probably transitional to the overlying upper Trolinger Spring Formation (Fig. 18c), which consisted of gray clay oxidized and mottled with yellowish-brown clay (Fig. 7c) away from the reducing environment of the spring and peat deposits.

A radiocarbon date on the top of unit c_2 shows it to be 51,000 ± 1500 B.P. (Fig. 19a), and juniper root wood on the basal contact of unit c_1 dated 48,900 ± 900 B.P. (QL-962) (Table 1), which suggests the tree was growing on top of unit c_2. The upper peat, being a mixture, may not be reliable for dating, but a date of greater than 35,320 (SMU-94) is consistent (Fig. 19b, line 0.0 North). Organic clay filling the nerve canal of a tusk from this level dated 31,130 ± 1550 B.P. (SMU-99). This appears to be too young considering that it is approximately the same date as samples from the peat (unit d_3) at Trolinger, which presumably postdates erosion of the Trolinger Spring Formation (Fig. 18a).

At Jones Spring, aggradation of Trolinger alluvium appears to have continued after deposition of the upper peat. No pronounced erosional contacts were seen in the overlying clay and the surface soil is more developed than in Boney Spring alluvium over Trolinger Spring.

In retrospect, Jones Spring appears to have burst forth at a time when about half of the thickness of the Trolinger Formation had been deposited. A stratigraphic break (unit c_2/c_1 contact) within the lower peat, while not very pronounced, appears to separate a significant faunal change which implies a time break during which there was no deposition of peat. This further implies that discharge of the spring was severely reduced or had stopped, as had aggradation of Trolinger alluvium, because the peat lens does not show any extensions of Trolinger clay into it as would be expected if the lower part of the lower peat (unit c_1) had been flooded before deposition of the upper part (unit c_2). The absence of a clearly defined contact and buried soil beyond the edges of the peat may be due to either the shortness of the break or to the blending effects of the saturated ground (gleying) since burial.

After an interval of unknown duration there was renewed peat accumulation, presumably in response to renewed spring discharge. Bison, in association with mammoth and mastodon at this time, were apparently a smaller species (cf. *antiquus*) than in the lower part of the lower peat (cf. *latifrons*) (Saunders, personal communication). Eventually, the peat settled and compacted as discharge declined and possibly stopped altogether. The absence of fossil pollen in the lower peat (unit c_1) suggests that it had dried out at one time, causing oxidation and destruction of pollen grains that must have been present. Unit c_1 and the implied hiatus of the c_2/c_1 contact possibly represents the Sangamonian interglacial (Haynes et al., 1983).

Eventually pressure on the underground water system increased to the point where it burst through the center of the lower peat lens and swept sand and gravel through it to establish a new conduit with sand feeders and capped by the upper peat with intermixed sand and clay. Bones, teeth, and fragments thereof were swept upward and redeposited in higher levels. The indicated period of increased hydrostatic pressure occurred between 38,000 and 51,000 B.P. and perhaps correlates with the Altonian Substage of Frye and Willman (1973). During a later phase of reduced discharge via subfeeders some mastodon bones worked their way downward and came to rest in the sand feeder. The final phase of Jones Spring appears to have been a small pond over the upper peat that eventually was buried under 2 m of Trolinger alluvium.

As mentioned earlier, there may have been an earlier discharge phase at Trolinger coinciding with one of those at Jones. The predominance of mammoth, bison, and horse remains in the conduit gravels at Trolinger compared to mastodon and musk oxen in the peat supports an earlier history of spring activity.

To clarify the stratigraphic relationship between Jones and Trolinger Springs, backhoe trenches (78A and B) were made from Jones Spring eastward across Jones Branch (Fig. 2). These provided a nearly continuous section between the two springs (Fig. 18) that reveals three alluvial cut-and-fill events following aggradation of the Trolinger Spring Formation. The first was the erosion preceding formation of the peat at Trolinger Spring followed by deposition of Boney Spring alluvium over the peat by floodwaters entering the cutoff meander. Between Jones Spring and Jones Branch the trenching revealed 3 m of the Rodgers Shelter Formation, which was cut-and-filled by 1.5 m of the Pippins Cemetery Formation (Fig. 18b). Terrace remnants of these units occur along Spring Creek, but, surprisingly, Rodgers alluvium does not extend up Spring Creek much beyond the confluence with Jones Branch, which has only about a third of the drainage area of Spring Creek. Trenches across the floodplain of Spring Creek upstream from the Breshears Valley revealed Boney Springs alluvium inset into Trolinger alluvium but no Rodgers or Pippins alluvium.

Kirby Spring

About 350 m west of Jones Spring is an elongated depression, measuring 27 m east-west and 5.5 m wide, containing a pond almost completely choked by cattails and other plants, including a willow tree (Fig. 4f). This spring, with a seepy outlet to Spring Creek 30 m north, we called Kirby after the last owner before purchase by the government. An 1845 land survey map is marked "Big Bone Diggings" at the site of Kirby Spring, which is probably the diggings referred to in a history of Benton County (Lay, 1876) as being worked by two brothers named Bradley (McMillan, 1976a).

Our test excavations in 1973 revealed that most of the deposits were thoroughly disturbed and no bones remained, but a peat layer did extend out from the spring on the eastern edge under a meter or more of gray clay. Radiocarbon dates show the peat to be more than 57,000 years old (Table 1), and fossil pollen indicates vegetation more xeric than that of today and unlike any other pollen profile in the Pleistocene deposits. These facts suggest that the peat at Kirby Spring, possibly the same age as the lower peat at Jones Spring, may be interglacial (Sangamonian?), because there is no known Wisconsinan substage with an essentially modern pollen spectrum.

Koch Spring

In 1971 we drained Koch Spring (Fig. 4e) by pumping from wells placed near it and drawing down the local water table. A north-south trench through the center of the spring exposed a brown (10YR 2/2) peat lens under 2.5 to 3 m of gray (5Y 6/2, 5/2) clay with chert gravel lenses, itself over interbedded clay, sand, and gravel (Fig. 13, V-1). The entire central part of the spring had been disturbed by Koch's and later excavations, and only a few fragments of bone and tusk were found in the mixed, pebbly clay and peat of the disturbed zone overlying the conduit area. This zone also contained a platform made of two layers of split wood rails, one layer at an angle to the other, that apparently served as a footing for previous excavators in removing fossil bones from what must have been a very muddy spring. Most of the split rails were good quality walnut, indicating its abundance then compared to today. A chert flake and a bifacial preform were found in the mixed layer and, like the artifacts found by Koch, probably derive from the late Archaic occupation in evidence around the spring (Wood, 1976).

As mentioned earlier, the peat lens at Koch Spring is between Boney Spring and Trolinger alluvium forming an 8.5-m terrace, a half meter of which is overbank deposition of Rodgers silt. Four radiocarbon dates from the peat range in age from

30,530 ± 200 B.P. from the top of the peat to 31,090 ± 150 B.P. from the bottom (Table 1). A date of 31,880 ± 1320 B.P. was on what appeared to be short fragments of yellow straw and vegetable fiber (Fig. 15f) that could be from the intestinal track of a mastodon or other large animal that fell into the spring.

Most of the peat has NAP and pine pollen which, with the radiocarbon dates, suggests correlation with the Plum Point Interstate. The top samples show a significant increase in spruce pollen much like that at the top of the younger peat (unit e) at Trolinger Spring (King, 1973). It is likely, therefore, that the alluvial clays (Fig. 13, units a_6-a_8) are Boney Spring Formation and represent aggradation during late Wisconsinan time. Wood from organic clay (unit b_1) under the peat is greater than 38,000 B.P., placing it and underlying units in the Trolinger Spring Formation.

Because of the severe disturbance of Koch Spring it could not be determined if multiple discharge events had occurred as at Jones and Trolinger springs or if discharge had at one time stopped altogether before final aggradation of the terrace (T-1a) at Koch Spring. As stated earlier, the terrace is covered by up to a meter of Rodgers silt, but the full thickness of Rodgers alluvium is not developed at the spring. Instead, the lower Pippins Cemetery Formation is inset against the Koch Terrace and contains organic deposits formed in a slough that was in part spring-fed (Fig. 13, V-5 and V-4). The organic sediments provided radiocarbon dates of between 840 and 620 B.P. for this episode of aggradation (Table 1).

Boney Spring

The 6.4-m terrace containing Boney Spring (Fig. 7c) 3 km northwest of the Breshears Valley (Fig. 2) is inset against a remnant of a higher terrace that appears to be Trolinger Spring Formation resting against the dolomite bluffs. The spring, which had a soft, bouncy vegetation mat around a small 'eye' of water in 1966, is in the Boney Spring Formation of terrace T-1b. An overbank facies of Rodgers alluvium extends over the springlaid deposits (Fig. 8b).

Taphonomic excavations of Boney Springs in 1971 uncovered mastodon bones representing 31 individuals (Fig. 7f) concentrated in a 0.7-m thick lens around a 2-m diameter feeder filled with calcareous coarse sand composed of granular tufa (Saunders, 1977a). The funnel-shaped feeder constricts to a diameter of about 1.2 m at the bottom, which is on gravel 5.5 m below the surface (Fig. 9). The blue chert gravel does not, however, extend up a conduit around the feeder as at Trolinger and Jones Springs. Five core holes along a line extending westward from east of the spring almost to the river show the basal gravel to extend all the way to the river with a thickness of 2 to 3 m (Fig. 8b). The basal gravel is overlain by bluish-gray (5Y 7/1, 5/1) and olive-gray (5Y 5/3, 4/1) clays with interbedded organic clayey sands and gravels and a brown pebbly peat (unit C_3) in the lower part of the section (Fig. 9). This entire section is one of vertical accretion. Fortunately, the occurrence of wood and organic samples allowed much of the succession to be radiocarbon dated (Table 1).

The base of unit B_{1b} provided dates of 24,460 ± 10,000 B.P. (Tx-1467) on peaty sand and 27,480 ± 1950 B.P. (Tx-1468) on soluble humates extracted therefrom. Neither date is satisfactory because of the imprecision of the first and possible secondary origin of the second, but both are consistent with respect to the date of 26,440 ± 1170 B.P. (Tx-1409) on wood from the second unit (C_3), a peat, above unit B_1. The contact between units B_2 and C_2-C_3 is sharp and appears to truncate unit B_2 (Fig. 9). If so, a significant hiatus is represented, and the lower units (B_1 and B_2) could be part of the Trolinger Spring Formation.

The upper part of the C_3 peat is dated at 22,730 ± 590 B.P. (Tx-1408) and a date of 28,330 ± 3140 B.P. (Tx-1407) on associated humates shows the mobility of soluble organic matter, probably a mixture, throughout the more permeable zones. Olive organic clay (unit C_4) overlying the peat has a basal date of 20,710 ± 530 B.P. (Tx-1479), a mid-section date of 21,380 ± 500 B.P. (Tx-1410), and a top date of 20,300 ± 470 B.P. (Tx-1474), all on wood. Two successive extractions of humates from the latter provides dates of 17,320 ± 1810 B.P. (Tx-1475) and 19,550 ± 1080 B.P. (Tx-1476), showing the mixed nature of the soluble fractions.

No suitable radiocarbon samples occurred in unit D, but several logs at the base of the bone bed (unit E_2) provided four dates all within one standard deviation of 16,500 B.P. (Table 1). Therefore, formation of the spring feeder occurred before this date but after 20,000 B.P. Correlation of this event with the Woodfordian Substage is supported by the pollen data (King, 1973).

The bone bed rests directly on the spruce logs and therefore postdates 16,500 B.P., as well as moss from the feeder dating 16,000 ± 400 B.P. (Tx-1629), which is probably derived from a live mat that existed around the 'eye.' Encrustation of the moss by calcium carbonate (tufa) suggests loss of carbon dioxide from the spring water either by evaporation or repeated wetting and drying by a fluctuating water level. Reduced discharge is indicated by either event, and the occurrence of divided tufa and moss fragments in the feeder indicate decay of the moss mat and break up of the tufa associated with a decline of discharge which correlates with deglaciation.

From careful excavation and detailed analysis of the bone bed, Saunders (1977a) concluded that the animals died after 16,200 B.P., during the time of stress caused by drought, and that many bones must have been exposed on the surface for some time, as indicated by their weathered condition. Unfortunately, the gray clay matrix of the bone bed is devoid of fossil pollen that would shed more light on the conditions that existed at the time of bone accumulation, but the nerve cavities of several tusks were filled with a dark-brown organic clay containing pollen of a mixed spruce and deciduous tree flora dated about 13,600 B.P. (Table 1). A similar pollen spectrum was found in a small lens of similar material within the bone bed (King, 1973). The filling of the tusks occurred after the bone bed had formed and after the tusks had become separated from the skulls, presumably upon weathering. The organic clay is apparently from a unit that formed over the bone bed during a time of moderate spring discharge following a period of reduced discharge.

These data imply a sequence of relative spring discharge varying from high discharge between 20,000 and 16,500 B.P., low discharge about 16,000 B.P. and moderate discharge until after 13,500 B.P., when the Boney Spring terrace (T-1b) was abandoned. Leighton (1960) called an "intraglacial" between Tazewell and Cary glacial substages the St. Charles substage (16,000 to 15,000 B.P.), and Mörner and Dreimanis (1973) described the Erie Interstade occurring in the same time interval between the Nissouri and Port Bruce Stades of the St. Lawrence region. As at Trolinger Spring, there does appear to be a correlation between glacial advance and spring discharge which reflects increased recharge of the aquifer supplying the spring.

A green clay (unit F) over the bone bed at Boney Spring contained a few flecks of scattered charcoal that dated 7290 ± 1900 B.P. (Tx-1466). This and overlying units (G, H, and I) appear to be overbank alluvium from the Pomme de Terre River modified by spring discharge. Unit I is a dark-brown clay incorporating a Woodland Indian site with three radiocarbon dates of approximately 1910 B.P. (King and McMillan, 1975) and a basal date of 4200 ± 140 B.P. (A-1076) (Table 1). A nuclear age radiocarbon value form the overlying unit J is on wood that is undoubtedly intrusive.

Phillips Spring

No sand-filled feeder or gravel conduit occurs at Phillips Spring, which consisted of a bed of moss over water in the center of a small depression 80 m from the edge of the Pomme de Terre River. Discharge was to the river via a small gully that may have been artificially deepened. As mentioned previously, the upper 2 m of strata around the spring contained numerous archaeological features and artifacts (Fig. 11).

The intense human occupation has disturbed the stratigraphy and obscured erosional contacts, but the spring is apparently due to a buried bulge in the basal gravel which brings the top of the aquifer closer to the surface at that point. The bulge is buried by the Rodgers Shelter Formation with a complex history of cutting and filling (Fig. 10a). It is quite possible that Phillips Spring may at one time have been a more typical spring, like Trolinger Spring, that was severely eroded during a major episode of degradation during which the feeder and overlying organic deposits were swept away, leaving only the conduit gravel as a lag deposit. A 4.5-m deep stratigraphic trench (Fig. 12) revealed organic clay in the basal gravels that yielded a radiocarbon date of 25,350 ± 440 B.P. (SMU-813). The soluble humate fraction was dated, so the age is imprecise, but it does support a pre-Rodgers Shelter Formation age for the basal gravel at Phillips Spring as does a pollen spectrum of 98-99% pine pollen (J. E. King, personal communication).

Such an origin would explain the absence of feeder sand; once swept away it would not be likely to reform because most of the sand had already been winnowed out of the gravel. Spring discharge would have kept pace with aggradation of Rodgers silt over the gravel bulge except when disturbed by one or more of the erosional episodes during the formation of terrace T-1c.

DEPOSITIONAL ENVIRONMENTS AND CORRELATIONS

From the detailed stratigraphy and radiocarbon dates presently available it is useful to compare top-stratum character, volume, deposition rates, and terrace heights for each fill over the past 100,000 years or so. These data are presented in Table 2, and Brakenridge (1981) has provided an independent interpretation. A hypothetical sequence of events is presented in Table 4.

The relatively uniform, fine-grained character of the Trolinger Spring Formation, with only minor tongues of interbedded colluvial gravels, all over a relatively thin bed of basal gravel, suggests deposition by a stream that was not meandering. It was more like the present river, although perhaps larger, which is made up of essentially straight segments with pools and riffles within the incised meanders of the ancestral river, and which aggrades by overbank deposition. Vertical accretion appears to have been the predominant depositional mode of aggradation up to the present day. From the stratigraphy at Jones and Trolinger Springs and the very low rate of peat deposition (net rate is 0.03 mm/yr), it appears that the Pomme de Terre was gradually aggrading throughout the vegetational change from juniper woodland 50,000 B.P. through open pine parkland up to 30,000 B.P. (King and Lindsay, 1976). Because of a lack of adequate dates on Kirby peat and basal Trolinger Formation, earlier depositional rates cannot be determined. The source of sediment within the drainage basin at this time is not known. Possible sources are slope washed loess, destruction of vertical accretion deposits of T-3, and stripping of upland soils including terra rosa and limestone residuum. Colors are generally those of reduced oxidation states that still prevail, except where oxidized under pedogenic conditions at the surface.

All radiocarbon dates older than 26,000 B.P. are on either peat itself or wood from one of the peat lenses. In evaluating a radiocarbon date on whole peat one must consider that the date is an average for the accumulation of fragments of moss, semi-aquatic plants, pondweeds, etc., over the time represented by the thickness sampled, usually 3 to 5 cm. At Boney Spring two dates on wood from the upper and lower parts of the peat (Fig. 9, unit C_3) provided a means of estimating an accumulation rate of 0.08 mm/yr, which amounts to 375 to 625 years for the thickness usually sampled. In the case of the Boney Spring peat this is less than the standard deviations of the radiocarbon dates and not, therefore, significant.

Radiocarbon dates on pieces of wood from peat deposits may not be representative of the host peat if they have intruded the peat by gravity (falling tree limb) or through animal trampling which would also churn up the peat and thereby disrupt any

TABLE 3. CORRELATION CHART

^{14}C Date*	Formation		Stratigraphic Unit	Symbol Figs. 2 & 3	Terrace Figs. 2 & 3	Midcontinent	E. Gt. Lakes	
190	PIPPINS CEMETERY		Upper Pippins alluvium	P_2	T-0			HOLOCENE
250								
430								
820			Lower Pippins alluvium	P_1				
1000	RODGERS SHELTER	Upper	Stratum 4 (G_2)**	R_6	T-1c			
1400								
1800								
2400			Stratum 4 (G_1)	R_5				
3600			Stratum 4 (G_1)	R_4				
4600								
5000		Middle						
			Stratum 3 (f)	R_3				
6300			Stratum 2 (E)	R_2				
8000		Lower						
10,500			Stratum a (A, B, C, D)##	R_1		GREATLAKEAN		
13,600	BONEY SPRING		Boney Spring tusk filling	BS	T-1b	Twocreekan	PORT BRUCE	LATE WISC.
16,500			Boney Spring bone bed (E)				Erie	
			Boney Spring alluvium (D)			WOODFORDIAN		
20,000			Boney Spring Alluvium (C_4)				NISSOURI	
22,000								
27,000			Boney Spring peat (C_3)			Farmdalian		
30,000	TROLINGER SPRING	Upper						M. WISCONSINAN
			Overbank deposition of Boney Spring Formation		T-1a			
31,000			Koch peat (b_2)				Plum Point	
32,000								
			Trolinger alluvium (upper)				CHERRYTREE	
			Trolinger peats (d_3) and				Port Talbot	
38,000			Jones upper peat (e_2)					
							GUILDWOOD	
		Middle	Jones Spring III			ALTONIAN		EARLY WISC.
			Trolinger alluvium (middle)					
>50,000			Jones peat (C_2)				St. Pierre	
			Jones Spring II					
							NICOLET	
		Lower	Jones peat (C_1) and Kirby peat			Sangamonian	Sangamonian	
			Jones Spring I and Kirby Spring	T	T-2			
						ILLINOIAN	ILLINOIAN	
>57,000			Trolinger alluvium (lower)					
			Slough deposit					
			Basal gravel					
			Breshears Valley Formation	B	T-3			

*In years B.P. and selected from Table 1.
**Unit designations.
#Stadials or Stades in upper case; interstadials or interstades in lower case.
##Units of Ahler (1973).

layering. A case in point is the modern radiocarbon date on a stick from the lower Pippins Cemetery Formation at Koch Spring, Trench V-5 (Fig. 13).

It is possible, if not probable, that some peat accumulates not in roughly horizontal layers but in inclined layers as flotsam is pushed by roiling water to the periphery of a spring pool or 'eye.' The radiocarbon dating of samples collected at Trolinger Spring in 1979 to test this hypothesis indicates that the more southerly parts of the peat cross section are older than the middle (Fig. 16b).

Except for the top sample or two from five of the ten pollen profiles in peats, most of the peat deposits represent relatively warm intervals of reduced spring discharge that probably correlate with interstadials in the glaciated areas to the north (Table 3). The peat (unit C_3) at Boney Spring is a convincing example because the pollen is dominated by NAP-pine indicating a pine parkland (King, 1973) and the radiocarbon dates are essentially the same as for the Farmdalian Substage (Frye and Willman, 1973).

Similar problems arise with correlation of the interstadial represented by the Koch peat. Radiocarbon dates from the periphery indicate formation during the Plum Point Interstadial, but the central portion of the peat lens could not be sampled because of removal by Koch's excavations.

The deposition rate of 0.6 mm/yr from Koch peat in Table 2 is based upon age estimates of ca. 2000 years for deposition of 1.2 m of peat. The age estimates are tenuous at best because of bulk sampling, but both Koch and Trolinger peats appear to have accumulated at an order of magnitude greater rate than Boney Spring peat C_3. At Koch Spring, alluvium older than the peat is Trolinger Spring Formation in a buried remnant of T-2 (Fig. 6b).

Slow aggradation of Trolinger Spring Formation (T-2) at ca. 0.09 mm/yr in the Breshears Valley ended before 38,000 B.P with only fluctuations of spring discharge and peat formation taking place thereafter, until slack water deposition of Boney Spring Formation occurred during Woodfordian time. The Trolinger Spring Formation appears to correlate with the Illinoian Glacial for the lower part, the Sangamonian Interglacial for the lower peat (unit C_1), and the Altonian Stade for the upper part with Trolinger peats (units d_3 and E) and Jones upper peat (unit e) correlating with the Port Talbot Interstadial (Table 3).

The very slow rate of deposition of the Trolinger Spring Formation within the Breshears Valley as compared to the Boney Spring Formation and younger formations outside of the valley is probably an indication that cutoff of the incised meander occurred before the beginning of deposition of the Trolinger top stratum in the valley. This is consistent with the interpretation offered earlier that the organic layer on top of the basal gravels (Fig. 18) is a slough related to abandonment of the valley by the Pomme de Terre River. All of the overlying fine-grained Trolinger alluvium confined to the Breshears Valley would, therefore, be from slack water deposition during high flood levels.

The Pomme de Terre began to aggrade (Boney Spring alluvium) again shortly before 26,000 B.P., during another period of open pine parkland (Table 4) that correlates more closely with the Farmdalian substage (27,000 to 22,000 B.P.) than with the Plum Point Interstade (32,000 to 22,000 B.P.). It continued to aggrade through the pollen-record change to spruce forest at ca. 0.3 mm/yr (Table 2), an order of magnitude faster than Trolinger alluvium. By 16,500 B.P. aggradation of T-1_b had ceased and a period of quasi-stability persisted for about 3,000 years, for which we have no pollen record. All we can say is that the forest cover changed from spruce dominance to a mixture of spruce and deciduous trees by 13,500 B.P. (Table 4). A lower water table during this period is inferred from the marked reduction of spring discharge, and Saunders (1977a) infers a drought that eliminated most of the Pleistocene megafauna in the Ozark Highland during the early part of this period.

Aggradation of Boney Springs alluvium took place during a time that Terasmae and Dreimanis (1976) call the Nissouri State and the last part of the preceding Plum Point Interstade. During the most intense part of the late Wisconsinan Glaciation, ca. 18,000 B.P., the Pomme de Terre floodplain was approaching a state of quasi-equilibrium approximately 6 m above the modern stream bed. The cessation of aggradation and the inferred dryness soon after 16,500 B.P. occurred at essentially the same time as the relatively dry Erie Interstade of about 15,500 B.P. (Mörner and Dreimanis, 1973). After 13,000 B.P. the river cut and eroded its banks for a period of about 2,000 years, during which time much of the Boney Spring alluvium must have been removed, as shown by its scarcity in the valley. (Therefore, original heights of T-1b are difficult to estimate). At the same time the Port Huron glacial advance followed by Two Creeks retreat occurred in the Great Lakes area to the north (Table 3).

Six epicycles of cutting and filling of brown clayey silt occurred during alluviation of the Rodgers Shelter Formation between 11,000 and 1400 B.P. (Table 4). Alluviation of about 6 m of the earliest Rodgers fill (R_1) occurred during the Greatlakean Substage and deglaciation that followed the Twocreekan Substage. The rate of deposition (2.0 mm/yr, Table 2) is an order of magnitude greater than for Boney Spring alluvium.

At Rodgers Shelter, deposition of talus beneath the protective overhang occurred between 11,000 and 10,000 B.P. or possibly somewhat earlier, but was a minor factor thereafter. The tabular dolomite clasts rest parallel to the talus slope, thus clearly defining the sequence of gradual accumulation (Ahler, 1976) and precluding deposition by stream flow, soil creep, or debris flow. A trench (76-A) on the T-1b terrace, to the west and outside of the shelter, revealed colluvial gravel that differs from that under the overhang, in being less flaggy and more dispersed (Kay, 1980). Therfore, the talus of Ahler's (1976) units A^2 and B^1 appears to be peculiar to the shelter, possibly the result of rock falls from the roof. If production was not the result of root action or efflorescence due to wetting and drying, but of frost action, which seems more likely, more freeze-thaw activity is implied for this period than any since then. What this means in terms of climate is less clear. Farrand (1975) has pointed out some of the problems with climatic interpretations of rockshelter deposits, but

TABLE 4. SEQUENCE OF GEOLOGICAL EVENTS AND VEGETATION CHANGES

	Event	Vegetation*	Approx. Age**
1	Deposition of Breshears Valley Formation and incision of meanders by point-bar deposition (T-3)		
2a	Degradation and abandonment of T-3		
2b	Low river discharge and red soil development		
3a	Increasing discharge		
3b	Cutoff of Breshears Valley meander		
3c	Organic slough deposits form in south arm, Breshears Valley		
3d	Vertical accretion of lower Trolinger Spring Formation (T-2)		
4a	Activation of Jones Spring (I) and possibly Trolinger and Kirby springs		
4b	Formation of older lower peat (unit C_1) at Jones Spring and peat at Kirby	Pine and oak Herbaceous prairie	
4c	Reduced discharge of Jones Spring		
5a	Increased discharge of Jones Spring (II)		
5b	Formation of younger lower peat (unit C_2) at Jones Spring	Pine and oak	>51,000
5c	Burial of Jones Spring by middle Trolinger Spring Formation (T-2)		
5d	Reactivation of Jones Spring (III) and formation of upper peat (unit e)		
5e	Formation of upper peat (unit e) at Jones Spring and Trolinger peats d_3 and e		38,000-32,000
5f	Burial of Jones Spring by upper Trolinger Spring Formation (T-2)		
5g	Soil formation, degradation and abandonment of T-2		
6a	Formation of peat at Koch Spring (unit b_2) and degradation	Pine parkland	32,000-31,000
6b	Vertical accretion over Koch peat of early Boney Spring Formation (T-1a) over buried remnants of T-2		
6c	Formation of peat C_3 at Boney Spring	Pine parkland	27,000
6d	Reduced discharge of Trolinger Spring and eventual desiccation and soil development		22,000
7a	Vertical accretion of Boney Spring Formation (T-1b)	Spruce forest	
7b	Activation of Boney Spring and occasional over bank deposition in Breshears Valley (unit f_2)		20,500
7c	Reduced discharge of Boney Spring (unit E) and accumulation of animal bones around the 'eye.'		16,500
7d	Accumulation of organic clay over bone bed at Boney Spring soil development	Mixed spruce and deciduous trees	13,500
7e	Degradation and abandonment of Boney Spring terrace (T-1b)		
8a	Vertical accretion of early Rodgers Formation (R_1)(terrace T-$1c_1$)(Strat. 1)	Oak-hickory	11,000-8,000
8b	Soil development followed by degradation		
8c	Vertical accretion of early Rodgers Formation (R_2)(terrace T-$1c_2$)(Strat. 2)	Oak-hickory	8,000-6,300
8d	Soil development followed by degradation		
8e	Vertical accretion of intermediate Rodgers Formation (R_3)(terrace T-$1c_3$)(Strat. 3 colluvium)		6,000-5,000
8f	Degradation		
8g	Vertical accretion of upper Rodgers Formation (R_4)(terrace T-$1c_4$)(Strat. 4)	Oak-hickory	4,600-2,900
8h	Degradation		
8i	Vertical accretion of upper Rodgers Formtion (R_5)(terrace T-$1c_5$)(Strat. 4)		2,400-1,800
8j	Degradation		
8k	Verticl accretion of upper Rodgers Formation (R_6)(terrace T-$1c_6$)(Strat. 4)		1,400-1,000
8l	Degradation and abandonment of Rodgers Terrace (T-1_c)		
9a	Vertical accretion of lower Pippins Formation (P_1)(terrace T-0_a)		820-430
9b	Degradation		
9c	Vertical accretion of upper Pippins Formation (P_2)(terrace T-0_b)		250-190
10	Discharge controlled by Pomme de Terre dam		10-0

*Dominant vegetation from King, 1973; King and Allen, 1977; King and Lindsay, 1976; King and McMillan, 1975.
**Radiocarbon dates in years B.P. (Before Present - 1950) from Table 1.

the necessity for data on rates of deposition should also be emphasized. A slow rate of alluviation can cause an increased percentage of colluvium, but this is not the case in Rodgers Shelter because the rate of deposition of the clayey silts (units A^1, B^2, and C) of Stratum 1 remained essentially the same during and after deposition of the talus (Table 2).

During the Hypsithermal interval, as redefined by Wright (1976), 5.2 m of Rodgers alluvium (R_2) was laid down, at a depositional rate of about 2.5 mm/yr at Phillips Spring, and buried the remnant of T-1_b at Boney Spring and R_1 at Rodgers Shelter, where the deposition rate for Stratum 2 was 2.0 mm/yr. The slower rate at which Stratum 2 was deposited in the shelter is consistent with the greater degree of pedogenesis apparent in it. The earliest part of R_2 deposition at Phillips Spring has the highest rate of deposition, 6.3 mm/yr, determined so far. As would be expected, the latest part was deposited at a slower rate, about 0.9 mm/yr. As the floodplain aggrades, the chances of overbank flooding and vertical accretion decreases (Wolman and Leopold, 1957, p. 97).

An erosional hiatus of no more than 1,000 years occurred during the transition from Hypsithermal to Neoglacial climates. Subsequent fills of Rodgers alluvium (R_3 through R_6) were deposited at rates 0.5 to 0.8 mm/yr through the transition from Hypsithermal to Neoglacial climates and during the Neoglacial as defined by Porter and Denton (1967). At Rodgers Shelter, overbank deposition may have reached Stratum 4 on rare occasions, but deposition of colluvium predominated during late Archaic and early Woodland occupations.

After an erosional hiatus of no more than 200 years, the first deposition of the lower Pippins Cemetery Formation (T-Oa), about 4 m of silty sand, occurred at an average rate of 5.0 mm/yr which was followed in less than a century by cutting and then filling of T-Ob at 1.0 mm/yr.

Alluviation of the Pomme de Terre over the past 50,000 years is separable into three periods that differ mainly in rates of aggradation and the character of the vertical accretion deposits. The Trolinger Spring and Boney Spring Formations are more clayey and chemically reduced than the Rodgers Shelter Formation and were laid down at much slower rates, although part of this could be attributed to the invisibility of epicycles of alluviation within each deposit, cut-and-fill contacts having been erased by the effects of time. For comparison, the net aggradation rate of Rodgers alluvium as a whole is 2.0 mm/yr, still greater than any of the earlier terraces, but less than half that for the Pippins Cemetery Formation (Table 2). Pippins alluvium is coarser and less oxidized than Rodgers (Brakenridge, 1979, 1981) and has the smallest volume and highest deposition rate.

Relationships of floodplain dimensions to channel dimensions are not known for the older terraces of the Pomme de Terre River, but if they were proportional to T-O and the modern channel, the relative size of past channels can be estimated by equating ratios of channel width to floodplain width:

$$\frac{C_p}{F_p} = \frac{C_r}{F_r} \qquad C_r = \frac{C_p\,F_r}{F_p}$$

where C_p is modern channel width, F_p is historic floodplain (Pippins) width, C_r is Rodgers channel width and F_r is the width of the Rodgers floodplain. The results (Table 2) suggest that there was little difference in channel dimensions during Rodgers and Boney Spring time, and probably, therefore, discharges were of the same order. The channel widths were approximately twice that of the modern Pomme de Terre, suggesting that average annual discharges were significantly greater than the 17.4 m^3/sec for the period of record (1912-1960) (U.S.G.S., 1961).

Channel width and discharge are even more tenuously estimated for the Trolinger Spring Formation because the deposits in the Breshears Valley are a special case of vertical accretion in a cutoff meander. With this in mind I estimate the channel width to have been half again wider than during Boney Spring or Rodgers time (Table 2), indicating greater discharge.

The increase in both grain size and rate of deposition from mid-Wisconsin time to the present is probably related to changing soil conditions over the approximately 2,000-km^2 catchment. The Trolinger and Boney Spring clays may have been derived from the filtering effect of grasses and other small plants on slope-washed loess. Pollen profiles consistently show 5-10% grass pollen, which would undoubtedly be considerably higher if the percentage constraints of pine and spruce were removed. The relatively high content of well-preserved pollen in both Koch and Boney Spring sediments indicates that reducing conditions have persisted since deposition. This is undoubtedly due to the sediments being within the zone of saturation and indicates a higher water table, which is consistent with the evidence of greater discharge from the springs in the past. Adjacent hillslopes may, therefore, have been moister, with more seeps, soil, and grass cover than during the time of deposition of the Rodgers Formation. Part of the soil cover probably consisted of loess, because the Breshears Valley is within an area of loess accumulation (Thorp and Smith, 1952), and loess has been traced into the area via drill-core sampling from thick outcrops north of the Missouri River (Johnson, personal communication). Brakenridge (1979, 1981) presents evidence, based partly upon clay mineralogy, that both the Boney Spring and early Rodgers Shelter Formations contain significant components of reworked loess, and an aeolian origin for silt in the early Rodgers Formation at the shelter was suggested by Ahler (1973, 1976).

No adequate pollen record exists for the early part of the Rodgers Shelter Formation, but there are records for the equivalent period from elsewhere in the region (King and Allen, 1977), and there is one grab (expediency) sample from the 7800 B.P. level (R_2) of the Rodgers terrace at Phillips Spring. These show a marked vegetational change to oak-hickory communities coincident with Rodgers alluviation. The relatively rapid appearance 10,000 to 11,000 years ago of the oak-hickory dominated forest was probably in part related to a reduction in the water table indicated by the oxidized sediment. This would likely have been accompanied by a reduction in seeps and plant cover on the valley sides, and Rodgers silt may reflect an increase in slopewash sediment. Brakenridge (1979, 1981) recognizes loess deposition during the time of deposition of the Rodgers Formation.

Stripping of a significant percentage of the plant substrate on slopes and the creation of unstable edaphic conditions may have been caused by more intense rain storms and, therefore, more concentrated runoff during Holocene time than had occurred during the Wisconsinan. On the basis of faunal evidence, McMillan (1976d) suggests that there was a gradual trend to increased grassland at the expense of forest from 8600 to 5000 B.P. and Purdue (1980) found clinal variation in some mammals at Rodgers Shelter that suggested a mesic early Holocene, less mesic middle Holocene, and essentially modern conditions in late Holocene time. The peak in numbers and variety of grassland vertebrates occurred at the base of Stratum 2 around 8000 B.P. when the river was beginning to aggrade (R_2). Fluctuations of these conditions and the removal of successive increments of slope soils in response to climate are probably the primary cause of the alluvial cut-and-fill cycles seen in the Rodgers Shelter Formation. By the end of Rodgers time most of the slope soils may have been removed and redeposited as alluvium. The resulting relatively rocky slopes would be even less effective in holding back runoff, which may account for the coarser character of the Pippins Cemetery Formation and the greater frequency of cut-and-fill cycles with time.

The basal gravels are essentially the same for all post-Breshears formations and are likely derived from reworking of T-3 gravels during each episode of alluviation. Unlike T-3, there is no evidence of lateral accretion in the strict sense (point-bar structure), but adequately long exposures are lacking. It is quite possible that the basal gravels are simply longitudinal pool and riffle depoists that were more or less reworked during flood stages of each cut-and-fill cycle. The present chert-gravel bedload appears to be inactive except during times of high discharge from the Pomme de Terre dam.

For Trolinger and Boney Spring deposits, source material color does not appear to have had much influence on the vertical accretion sediments because the red colors of T-3 did not carry over, or if they did they were subsequently reduced. The reddish-browns of Rodgers alluvium are in part due to oxidation as indicated by the gray color, reduced state, in the saturated zone of Rodgers Shelter and by the absence of fossil pollen in the aerated zone, as indicated by many attempts to recover it. The brownish-gray color of Pippins alluvium appears to be due to a combination of its freshness, organic content, and, in some cases, dispersed charcoal, that may indicate an upsurge of forest fires possibly set by Indians to improve the forest for game (King, 1982).

Surface colors of the chert gravels clearly reflect oxidation states. In most cases those at the base of the Trolinger Spring and Boney Spring Formations and in their spring conduits are dark bluish-gray (5B 4/1 on the Munsell gley color chart), indicating reducing conditions. On the other hand, those of the Breshears Valley Formation and basal Rodgers Shelter and Pippins Cemetery Formations are typically oxidized to a strong brown color (7.5YR 5/7), much like the present bedload. Very likely, all of the bluish-gray gravels were derived from the reworking of brown Breshears gravel. It appears that reduction occurs once the gravels become inactive and isolated within anaerobic waters (older ground water?), but the color change from brown to gray apparently takes thousands of years to complete. The reverse process is apparently much faster because the modern bedload, with few exceptions, is oxidized. The few exceptions are where chert pebbles are mottled brown and blueish-gray and appear to represent a transitional stage.

The best example of the oxidation-reduction phenomenon is at Phillips Spring where the deep stratigraphic trench (Fig. 12) exposed brown chert channel gravel in the 8000 year old channel cut into bluish-gray chert gravel about 25,000 years old. It is not known whether the brown gravel was swept over the gray gravel or was a result of the reworking and oxidation of the older gravel. Perhaps it was a combination of both.

PALEOCLIMATES, PALEOHYDROLOGY, AND ALLUVIAL PROCESSES

There appear to be approximate correlations of the alluvial events of the Pomme de Terre Valley to climatic events inferred from the glacial record. Unless tectonics are involved, maximum spring discharge appears to correlate with glacial maxima. Spring discharge partly follows and partly coincides with vertical accretion, implying contemporaneity with more and/or larger overbank floods. The peat deposits at springs form during periods of reduced discharge and less mesic climate corresponding to interstadials, but as Table 3 shows, precise correlations are equivocal. More mesic pollen at the top of some peats may indicate intrusive contamination or renewed accumulation of peat, but a change to more mesic conditions is indicated in either case. The gravels and few remains of fine-grained facies of the Breshears Formation (T-3) represent a more vigorous river than at any stage since their deposition. They may be pre-Illionian in age. The strong red soil development on the Breshears Valley Formation can logically be correlated with the typically red Sangamon soil in well-drained sites of the Midcontinent. The peat at Kirby Spring and the lower peat (unit C_1) at Jones Spring are correlated with the Sangamonian Interglacial, and aggradation of the lower Trolinger Spring Formation with the Illionian Glacial. Stage I of Jones Spring discharge occurred during this time (Table 3).

Perhaps deposition of the upper Trolinger Formation began early during the Nicolet Stade of Terasmae and Dreimanis (1976) with Jones Spring, Stage II occurring late and formation of the intermediate peat (C_2) during the St. Pierre Interstade of Dreimanis and Goldthwait (1973) (Table 3, B). Carrying this interpretation further, the rest of Trolinger aggradation and Jones Spring II would have occurred during the Guildwood Stade with the Jones peat (unit e_2) and peats d_3 and e of Trolinger Spring corresponding to Port Talbot I and II Interstades. There is no named stade between these two interstades unless it is represented by the Dunwich till of Dreimanis and Goldthwait (1973). A contact between Trolinger peats d_3 and e may, perhaps, correlate with renewed spring discharge at this time, but the correlation is admittedly tenuous.

The Koch peat appears to fall within the Plum Point Interstade, so the preceding inferred period of higher spring discharge may correlate with the Cherrytree Stade. The correlations of the Boney Spring stratigraphy with the Nissouri Stade and subsequent deglaciation is more convincing, as already stated.

While deglaciation was taking place there was a general lowering of water tables in the Ozarks as well as elsewhere. The eventual effect of this was the reduction of soil moisture and seeps on hillslopes, with the result that the character of slope soils changed. This change in the substrate for plant growth led to more xeric vegetation and more unstable slope cover. Perhaps this, in combination with a climatic change to more intense storms, increased slope erosion and sediment load. Early Rodgers colluvium at the shelter appears to have been partly ablation slabs (eboulis) from the roof, presumably caused by more intense freeze-thaw activity than at anytime since. This was followed by relatively little colluvial deposition until, during Stratum 2 time, more colluvial dolomite and chert gravel appears in the sediments and reaches a maximum during Stratum 3, which may represent an alluvial fan or mud flow formed during a short interval. Colluvial processes remained dominant during Stratum 4 time. This sequence appears to reflect increasing slope-wash alluviation and hillslope erosion, with the result that more bare rock is exposed today than in the past, yielding more coarse colluvial gravel, interbedded with fine-grained alluvium.

Brakenridge (1981) suggests that floodplain stability 20,000 to 13,500 B.P. (Boney Spring Formation) and 7500 to 5000 B.P.

(middle Rodgers Shelter Formation) was caused by intensified zonal circulation of the upper atmosphere (Knox, 1976), which resulted in less frequent large floods and more frequent small floods (Knox and others, 1975). The periods referred to are actually times of aggradation of vertical accretion deposits (Table 2), not stability, which came at the end of the periods. By less frequent large floods Brakenridge apparently means less overbank deposition. Conceivably this could reduce vertical accretion to a point of stability.

Brakenridge further suggests that downcutting is caused by intensified meridional circulation, resulting in increased stream power (Bull, 1979) and more frequent floods, but the character and size of the floods is not stated. It is well-established that overbank floods can either add to or erode the surface of floodplains, but vertical accretion is the net effect of overbank floods.

From the alluvial history of the Pomme de Terre presented here, it is clear that degradation has been mostly by bank erosion and not channel deepening. It is conceivable, therefore, that a reduction of effective floods to less than bankfull stage could result in termination of floodplain aggradation, although erosion of the stream banks would eventually remove large parts of the terraces, especially if the sediment yield from drainage basins has been reduced by prior stripping of the bedrock slopes. Perhaps at this stage lateral accretion would become the predominant fluvial process if most of the floodplain (top stratum) were to be removed, but the evidence shows little time at this stage before the river reverts to an aggradational mode. Under these hypothetical considerations, degradation appears to have been produced by lower discharges than was aggradation, a situation not uncommon in arroyos.

Under what circumstances did the regimen reverse and change from degradation to aggradation? Obviously more sediment and more frequent overbank floods are required to build a new floodplain by vertical accretion. Sediment could come from lateral erosion of older terraces and from slope wash. Older terraces have undergone more erosion during the degradational stage than during aggradation, so slope wash would appear to be the predominant source. Widespread loess deposition was greatest during the Pleistocene and coincided with glacial advances (Frye and others, 1968). Weathering of bedrock and soils thereon appears to have been a lesser source of slope sediment than loess deposition during the late Quaternary. If loess blanketed the uplands, more intense rains would be required to wash it to the streams. The greater the effect of vegetation in protecting slopes, the more intense the storms would have to be to overcome the effect. These factors tend to indicate that floodplain aggradation of the Pomme de Terre by vertical accretion during the late Pleistocene was caused by a change to more intensive storms occurring during a time of loess accumulation on adjacent slopes. Aggradation would continue as long as overbank flooding occurred and the slopes yielded sediment, but at a diminishing rate as the height of the floodplain rose and/or as the sediment yield decreased by either stripping to bedrock or growth of more effective vegetative cover (Schumm, 1965).

According to this reasoning there appears to have been stability within the Boney Spring Formation (Fig. 9) between 26,000 and 22,000 B.P., the Farmdalian Substage (interstadial) when unit C_3 peat accumulated at a rate of only 0.08 mm/yr (Table 2). There was net aggradation (0.3 mm/yr, Table 2) during the Woodfordian Substage (units C_4-E_2), with hydrostatic pressure reaching a maximum sometime before 16,500 B.P. at Boney Spring when the feeder formed. These events indicate that overbank flooding and vertical accretion coincided with increasing hydrostatic pressure on the basal gravel aquifer, which indicates that ground water recharge was increasing during glaciation.

Formation of the Boney Spring feeder appears to have coincided, within a millennium or so, with the Late Wisconsin maximum, because by 16,500 B.P. discharge had declined, and the floodplain remained stable during the Erie Interstadial. Stability continued during the final phases of Late Wisconsin deglaciation until a degradational mode set in between 13,500 and 11,500 B.P. centered in the Two Creeks Interstadial.

It isn't clear whether the indicated increased hydrostatic pressure at springs and discharge of the Pomme de Terre occurred because of increased precipitation or reduced mean annual temperature (Brakenridge, 1978, 1980). The slope washing and overbank flooding would appear to require more than just a decrease in temperature, but this does not necessarily mean there had to be a significant increase in rainfall. A change in rainfall intensity or an increase in the frequency of the more intense storms could produce the same result (Knox, 1976).

Unconsidered so far is the relative effect of runoff versus groundwater effluent to stream discharge. Groundwater entering the stream via underwater springs and through the basal gravels will have the same power to erode as runoff without sediment. It appears, therefore, that bank erosion could occur as a result of groundwater discharge continuing after (lagging) the period of maximum recharge, in other words during the transition from a glacial stade to an interstade. This may have been augmented by a reduction in sediment yield upon a lessening of loessial fallout.

Minimum discharge would be attained through reduced precipitation and/or increased mean annual temperature when groundwater discharge had reached a minimum. Both recorded and oral history show that the Pomme de Terre ceased to flow for several days twice during droughts of the past 50 years (Brakenridge, 1979). At these times it was an influent stream. If the influent mode became a semi-permanent condition it would require more than reduced mean annual temperature to return the discharge to the proportions indicated by the Pleistocene alluvial terraces and periods of maximum spring discharge. It appears that some degree of increased precipitation is required as well.

The difference in the lithology of late Pleistocene versus Holocene alluvium and cut-and-fill frequency is probably due to several factors, not the least of which is reduced discharge. With the end of glaciation there was a reduction of loess deposition with only a small amount, the Bignell loess (Frye and others, 1968), occurring in the earliest Holocene and probably coeval with the Greatlakean Substage (Evenson and others, 1976). With diminishing loess deposition during early Rodgers time, sediment on the slopes was being removed faster than it was being replenished. The greater deposition rates and the coarser sediment of Rodgers alluvium compared to the earlier ones may have been the result of a less effective vegetative filter brought about by the establishment of the deciduous forest and open glades during the Holocene.

The increase in the frequency of cut-and-fill cycles during the Holocene is also not clearly understood. As mentioned before, Brakenridge (1981) attributes it to Holocene climatic change and applies the model of Knox (1976) in attributing more intense, and therefore erosive, rains to meridional flow of the upper atmospheric circulation pattern as the explanation, but I believe it is equivocal whether the net result would be degradation or vertical accretion along the Pomme de Terre.

Another factor might be human utilization of the floodplains. Natural floodplains were probably covered by dense growths of vegetation that reduced the velocity of overbank flow and promoted sedimentation. Disruption of this cover by fire or artificial clearing would lower the threshold for erosion.

As previously mentioned, dispersed charcoal as well as charcoal from archaeological hearths was found in several of the test trenches in Rodgers alluvium, which helped establish the radiocarbon chronology. Charcoal and archaeologically produced chert flakes were found at 2, 6, and 8 m below the surface in Rodgers alluvium at core hole No. 15 at the mouth of the Breshears Valley (Fig. 2). Fires were often deliberately set to clear away brush and other vegetation and to drive game. The earliest evidence of man in the Pomme de Terre Valley is the 10,500 B.P. occupation of Rodgers Shelter by hunters, during aggradation of early Rodgers alluvium (R_1). Foraging continued to be the main subsistence base through the Archaic period, during which five erosional episodes occurred that may have been triggered by floodplain fires during periods of climatic stress and less overbank deposition such that stream power exceeded the lowered threshold for net erosion. By late Archaic time agriculture was practiced at the Phillips Spring site (Kay, 1979) and thereafter the clearing of floodplain land for farming probably increased, further lowering the threshold for erosion. Charcoal is even more abundant in Pippins alluvium, suggesting more intensive human use of floodplains.

CONCLUSIONS

The once-meandering ancestral Pomme de Terre River has, since perhaps Illinoian time, undergone at least four major cycles of degradation and vertical accretion represented by four terraces confined to its valley incised into cherty Ordovician dolomites. During Holocene time there have been eight epicycles of cutting and filling. Chert gravels underlying all of the vertical accretion deposits may be relict lateral accretion deposits that were reactivated at flood stage only when they coincided with the active channel, where they formed pools and riffles. The basal gravels form an aquifer feeding several artesian springs that have erupted to the surface of floodplains at times of high hydrostatic pressure that appear to correlate with glacial maxima. Floodplain aggradation appears to take place under mesic conditions coinciding with glacial advance in more northerly regions, and stability of the floodplain surface with the onset of deglaciation. Peat deposits over spring conduits and feeders reflect reduced discharge and contain pollen, plant remains, and animal bones, indicating less mesic interstadial conditions.

Sediment yield accompanying aggradation may be from loess derived from glacial outwash elsewhere and washed from slopes in the drainage basin. Slope vegetation acted as a filter on slope wash, allowing finer fractions of sediment to charge the streams. Vertical accretion occurred during periods when overbank floods caused net aggradation. Degradation occurred by bank erosion during times of reduced overbank flooding.

Generally coarser Holocene alluvium may have been due to less effective vegetative filtering as well as the reduction of loess deposition and slope washing of the coarser fractions left after the Pleistocene winnowing of fines. The increased frequency of cut-and-fill epicycles during the Holocene could be due to a combination of factors, such as increased frequency of intense storms, reduced sediment yield from increased denudation of bedrock slopes, and reduction of floodplain vegetation by human activity. For whatever reason, there is an approximate correlation between these epicycles in the Ozark Highlands and the western United States. At least for the early Holocene there appears to be approximate correlation to the few documented sequences in the East (Coe, 1964; Gardner, 1977) and Midwest (Butzer, 1977; Knox and others, 1975) as well as in Europe (Brakenridge, 1980).

As in many other stream valleys in the United States, there is a marked difference in the color, texture, and amount of late Pleistocene alluvium compared to Holocene alluvium; a pronounced interval of degradation, occurring between 13,000 and 11,000 B.P., separates them.

Paleoindians first appear at Rodgers Shelter 10,500 years ago and the occurrence of a few fluted point surface finds indicates occupation of the Pomme de Terre area as early as 11,000 B.P. In spite of more abundant plants and game animals, there is no evidence of man before this. Because of Koch's claim for the association of artifacts with mastadon bones and because we were scientifically excavating deposits with the best possible prospects of having Paleoindian artifacts, all personnel involved over 14 consecutive field seasons were constantly on the lookout for artifacts *in situ* in deposits predating the Rodgers Shelter Formation (Wood, 1976). None were found.

This project provided an excellent opportunity to study the complex stratigraphy of alluvial springs and to compare these to others in the western United States such as Tule Springs, Nevada (Haynes, 1967), U. P. Mammoth site, Wyoming (Haynes, n.d.), Blackwater Draw, New Mexico (Haynes and Agogino, 1966), and Murray Springs, Arizona (Haynes, 1980). The sand feeders are common to all of these, but the Pomme de Terre springs are the only ones in my experience with concentric conduit gravels.

Basal alluvial gravels are common to all of the Pomme de Terre springs but in the western examples they are only found at the type Clovis site in Blackwater Draw (Haynes and Agogino, 1966). The western springs may not have developed concentric conduit gravels because of 1) the lack of sufficient hydrostatic pressure in the relatively flat terrain or 2) the consolidation (*calichefication*) of the overlying strata or 3) possibly the rapid tectonic formation of a fissure.

A common factor in all of these late Pleistocene springs is the preservation of the remains of Rancholabrean "big game" animals. The springs are natural animal traps. Only the youngest of these, postdating 12,000 B.P., have evidence of human predation.

In spite of possibly more detailed data on late Quaternary chronostratigraphy than anywhere else in the unglaciated midcontinent, we still do not have enough pertinent facts about the interrelationships of climate, vegetation, and fluvial processes on the lower Pomme de Terre to make these conclusions any more than working hypotheses. Further understanding may be gained by modeling the Pomme de Terre fluvial systems; hopefully someone will try this approach. In the meantime, we are processing a few additional radiocarbon samples that will help provide a more precise chronology for the mid-Wisconsinan part of the succession.

Late Quaternary stratigraphy investigations in the Ozark Highland and eastern United States are seriously hampered by the dense vegetative cover compared to the western part of the country. Natural exposures are thus limited and commonly inadequate. Core data help, but mostly in testing questions raised by stratigraphic mapping from artificial excavations. The backhoe is undoubtedly the most useful tool yet devised for stratigraphic mapping of alluvium, but the trenches can be dangerous, especially where shallow ground water is reached. The best policy in such circumstances is to remove one wall of the trench. Bracing of walls with stulls, cribbing, etc. is unsatisfactory because such measures are not only time consuming and very inconvenient but are likely to hide the often subtle contacts one is looking for.

Since its impoundment in 1978, fluvial processes are no longer active along the lower Pomme de Terre. Instead we can look forward to gradual sedimentation of the Harry S. Truman Lake by deltaic accretion and lacustrine muds until a time when future generations find they have a giant mudflat. This might once again bring agriculture to the lower Pomme de Terre to help feed what might then be an overpopulated world of hungry people.

REFERENCES CITED

Ahler, S. A.

1973 : Post-Pleistocene depositional change at Rodgers Shelter, Missouri. *Plains Anthropologist* 18:1-26.

1976 : Sedimentary processes at Rodgers Shelter, Missouri. In *Prehistoric man and his environments: a case study in the Ozark Highland*. edited by W. R. Wood and R. B. McMillan, pp. 123–139. Academic Press, New York.

Brakenridge, G. R.

1978 : Evidence for a cold, dry, full-glacial climate in the American Southwest. *Quaternary Research* 9:22-40.

1979 : *The impact of climatic change on floodplain sedimentation, soil formation, and eolian activity in southern Missouri.* [M.S. thesis], Department of Geological Sciences, University of Arizona, Tucson, 55 pp.

1980 : Widespread episodes of stream erosion during the Holocene and their climatic cause. *Nature* 283:655–656.

1981 : Late Quaternary floodplain sedimentation along the Pomme de Terre River, southern Missouri. *Quaternary Research* 15:62–76.

Bull, W. B.

1979 : Threshold of critical power in streams. *Geological Society of America Bulletin* 90:453–464.

Butzer, K. W.

1977 : Geomorphology of the Lower Illinois Valley as a spatial-temporal context for the Koster Archaic Site. *Illinois State Museum Reports of Investigations* 34, 60 pp.

Chomko, S. A.

1976 : The Phillips Spring Site, 23H1216: Harry S. Truman Reservoir, Missouri. *Univ. Missouri Dept. Anthro. Report for National Park Service USDA,* 124 pp.

1978 : Phillips Spring, 23H1216: a multicomponent site in the western Missouri Ozarks. *Plains Anthropologist* 23:235–255.

Coe, J. L.

1964 : The formative cultures of the Carolina Piedmont. *Transactions of the American Philosophical Society* 54:130.

Dreimanis, A., and R. P. Goldthwait

1973 : Wisconsin glaciation in the Huron, Erie, and Ontario Lobes. In *The Wisconsinan Stage,* edited by R. F. Black, R. P. Goldthwait, and H. B. Willman. *Geological Society of America Memoir* 136:71–105.

Evenson, E. B., W. R. Farrand, D. M. Michelson, D. F. Eschman, and L. J. Maher

1976 : Greatlakean Substage: a replacement for Valderan in the Lake Michigan Basin. *Quaternary Research* 6:411–424.

Farrand, W. R.

1975 : Sediment analysis of a prehistoric rockshelter: the Abri Pataud. *Quaternary Research* 5:1–26.

Fenneman, N. H.

1938 : *Physiography of eastern United States.* McGraw Hill, New York.

Frye, J. C., and H. B. Willman

1973 : Wisconsinan climatic history interpreted from Lake Michigan Lobe deposits and soils. *Geological Society of America Memoir* 136:135–152.

Frye, J. C., H. B. Willman, and H. D. Glass

1968 : Correlation of Midwestern loesses with the glacial succession. In *Loess and related eolian deposits of the world,* edited by C. B. Schultz and J. C. Frye. Proceedings VII INQUA Congress 12:3–21. Univ. Neb. Press, 369 pp.

Gardner, W. M.

1977 : Flint Run paleo-Indian complex and its implications for eastern North American prehistory. In *Amerinds and their Paleoenvironments in northeastern North America,* edited by W. S. Newman and B. Salwen. *Annals of the New York Academy of Science* 288:257–263.

Haynes, C. V., Jr.

1967 : Quaternary geology of the Tule Springs area, Clark County, Nevada. *Nevada State Museum Anthropology Paper* 13:15–104.

1968 : Geochronology of Late-Quaternary alluvium. In *Means of correlation of Quaternary successions,* edited by B. Morrison and H. E. Wright, Proceedings VII INQUA Congress 8:591–631. Univ. Utah Press, Salt Lake City.

1976 : Late Quaternary geology of the lower Pomme de Terre Valley. In *Prehistoric man and his environments: a case study in the Ozark Highland,* edited by W. R. Wood and R. B. McMillan, pp. 47–61, Academic Press, New York.

1978 : *Geochronology of the lower Pomme de Terre River.* Paper presented 43rd Annual Meeting of the Society for American Archaeology, Tucson.

1980 : Archaeological investigations at the Murray Springs Clovis site, Arizona, 1971. *National Geographic Society Research Reports* 12:347–353.

1983 : Report on geochronological investigations in the Harry S. Truman Reservoir area, Benton and Hickory Counties, Missouri. *Cultural Resources Survey Harry S. Truman Dam and Reservoir Project* 10:25–34.

n.d. : Geochronology of the U.P. Mammoth Site, Wyoming. Unpublished manuscript.

Haynes, C. V., Jr. and G. A. Agognino

1966 : Prehistoric springs and geochronology of the Clovis Site, New Mexico. *American Antiquity* 31:812–821.

Haynes, C. V., Jr., M. Stuiver, H. Haas, J. E. King, F. B. King, and J. J. Saunders

1983 : Mid-Wisconsinan radiocarbon dates from mastodon- and mammoth-bearing springs, Ozark Highland, Missouri. *Radiocarbon* 25:381–391.

Johnson, D. L.

1983 : Soils and soil-geomorphic investigations in the lower Pomme de Terre Valley. *Cultural Resources Survey Harry S. Truman Dam and Reservoir Project* 10:63–143.

Kay, Marvin

1979 : Phillip S. Spring, Missouri: report of the 1978 investigations. *Report for the U.S. Army Corps of Engineers, Kansas City District, under contract No. DACW41-76-C-0011 to the Illinois State Museum.*

1980 : Stratigraphic studies at Rodgers Shelter. In *Holocene adaptations within the lower Pomme de Terre River Valley, Missouri,* edited by M. Kay, pp. 81–106. Report to the U.S. Army Corps of Engineers, Kansas City District, under contract No. DACW41-76-C-0011 to the Illinois State Museum.

King, F. B.

1982 : Vegetational reconstruction and plant resources prediction. In *Holocene adaptations within the lower Pomme de Terre River Valley, Missouri,* edited by M. Kay, pp. 9–61. Report to the U.S. Army Corps of Engineers, Kansas City District, under contract No. DACW41-76-C-0011 to the Illinois State Museum.

King, F. B., and R. B. McMillan

1975 : Plant remains from a Woodland storage pit, Boney Spring, Missouri. *Plains Anthropologist* 20:111–115.

King, J. E.

1973 : Late Pleistocene palynology and biogeography of the western Missouri Ozarks. *Ecological Monographs* 43:539–565.

King, J. E., and W. H. Allen, Jr.

1977 : A Holocene vegetation record from the Mississippi River Valley, southeastern Missouri. *Quaternary Research* 8:307–323.

King, J. E., and E. H., Lindsay

1976 : Late Quaternary biotic records from spring deposits in western Missouri. In *Prehistoric man and his environments: a case study in the Ozark Highland,* edited by W. R. Wood and R. B. McMillan, pp. 65–78, Academic Press.

Knox, J. C.

1976 : Concept of the graded stream. In *Theories of landform development,* edited by W. B. Melhorn and R. C. Flemal, pp. 169–198. State University of New York Publications in Geomorphology, Binghamton, New York.

Knox, J. C., P. J. Bartlein, K. K. Hirschboeck, and R. J. Muckenhirn

1975 : The response of floods and sediment yields to climate variation and land use in the upper Mississippi Valley. *University of Wisconsin Institute for Environmental Studies Report,* No. 52, 75 pp. Madison, Wisconsin.

Lay, J. H.

1876 : *A sketch of the history of Benton County, Missouri.* Winchell and Ebert Printing and Lithograph Company, Hannibal, Missouri.

Leighton, M. M.

1960 : The classification of the Wisconsin glacial stage of North Central United States. *Journal of Geology* 68:529–552.

McKinney, Curtis R.

1979 : Implications of high U^{234} springs for dating Early Man. *Geological Society of America, Abstracts with Programs* 11:454.

McMillan, R. B.

1971 : *Biophysical change and cultural adaptation at Rodgers Shelter, Missouri.* [Ph.D. dissertation], University of Colorado, Boulder.

1976a: Man and mastodon: a review of Koch's 1840 Pomme de Terre expeditions. In *Prehistoric man and his environment: a case study in the Ozark Highland,* edited by W. R. Wood and R. B. McMillan, pp. 82–96, Academic Press, New York.

1976b: The Pomme de Terre study locality: its settings. In *Prehistoric man and his environments: a case study in the Ozark Highland,* edited by W. R. Wood and R. B. McMillan, pp. 14–44, Academic Press, New York.

1976c: Rodgers Shelter: a record of cultural and environmental changes. In *Prehistoric man and his environments: a case study in the Ozark Highland,* edited by W. R. Wood and R. B. McMillan, pp. 111–122, Academic Press, New York.

1976d: The dynamics of cultural and environmental change at Rodgers Shelter, Missouri. In *Prehistoric man and his environments: a case study in the Ozark Highland,* edited by W. R. Wood and R. B. McMillan, pp. 212–232, Academic Press, New York.

Mehringer, P. J., Jr., C. E. Schweger, W. R. Wood, and R. B. McMillan

1968 : Late-Pleistocene boreal forest in the western Ozark Highlands. *Ecology* 49:567–568.

Mörner, N. A., and A. Dreimanis

1973 : The Erie Interstade. In *The Wisconsinan stage,* edited by R. F. Black, R. P. Goldthwait, and H. B. Willman, pp. 107–134, *Geological Society of America Memoir* 136.

Porter, S. C. and G. G. Denton

1967 : Chronology of neoglaciation in the North American Cordillera. *American Journal of Science* 265:177–210.

Purdue, J. R.

1980 : Clinal variation of some mammals during the Holocene in Missouri. *Quaternary Research* 13:242–258.

Robinson, C. K., and M. Kay

1980 : Phillips Spring excavations and archaeology. In *Holocene adaptations within the lower Pomme de Terre River Valley, Missouri,* edited by M. Kay, pp. 623–682. Report for U.S. Army Corps of Engineers, Kansas City District, under contract No. DACW41-76-C-0011 to the Illinois State Museum.

Saunders, J. J.

1977a: Late Pleistocene vertebrates of the western Ozark Highlands, Missouri. *Illinois State Museum, Reports of Investigations No. 33,* 118 pp.

1977b: Jones Spring Excavation 1977, field report, summary of synthesis to date. *Unpublished report for Department of the Army, Corps of Engineers, Kansas City, Missouri,* 10 pp.

Schumm, S. A.

1965 : Paleohydrology. In *The Quaternary of the United States,* edited by H. E. Wright and D. G. Frey, pp. 783–794. Princeton University Press, Princeton, New Jersey.

Szabo, B. J.

1982 : Extreme fractionation of $^{234}U/^{238}U$ and $^{230}Th/^{234}U$ in spring waters, sediments, and fossils at the Pomme de Terre Valley, southwestern Missouri. *Geochimica et Cosmochimica Acta* 46:1675–1679.

Thorp, J., and H.T.U. Smith

1952 : Pleistocene eolian deposits of the United States, Alaska, and parts of Canada (ed. map). *Geological Society of America.*

Terasmae, J., and A. Dreimanis

1976 : Quaternary stratigraphy of southern Ontario. In *Quaternary stratigraphy of North America,* edited by W. C. Mahaney, pp. 51–63. Dowden,

Huchinson and Ross.
U.S. Army Corps of Engineers
1975 : Truman Dam and Reservoir, Missouri. Map with explanations, Kansas City, Missouri.
U.S. Geological Survey
1961 : Surface water supply of the U.S. to 1960. Part 6-B: Missouri River Basin below Sioux City, Iowa. *Water Supply Paper* 2119: 899 pp.
West, J.
1945 : *Plainville U.S.A.* Columbia Univ. Press, 239 pp.
Wolman, M. G., and L. B. Leopold
1957 : River flood plains: some observations on their formation. *U.S. Geological Survey Professional Paper* 282-C:85-109.
Wood, W. R.
1976 : Archaeological investigations at the Pomme de Terre springs. In *Prehistoric man and his environments: a case study in the Ozark Highland,* edited by W. R. Wood and R. B. McMillan, pp. 97–107. Academic Press, New York.
Wood, W. R., and R. B. McMillan (editors).
1976 : *Prehistoric man and his environments: a case study in the Ozark Highland,* 271 pp. Academic Press, New York.
Wright, H. E., Jr.
1976 : The dynamic nature of Holocene vegetation: a problem in paleoclimatology, biogeography, and stratigraphic nomenclature. *Quaternary Research* 6:581–596.

MANUSCRIPT ACCEPTED BY THE SOCIETY MARCH 5, 1985

Typeset by WESType Publishing Services, Inc., Boulder, Colorado
Printed in U.S.A. by Malloy Lithographing, Inc., Ann Arbor, Michigan